Statistics for Social and Behavioral Sciences

Statistics for Social and Behavioral Sciences (SSBS) includes monographs and advanced textbooks relating to education, psychology, sociology, political science, public policy, and law.

Gerhard Tutz

A Short Guide to Item Response Theory Models

Springer

Gerhard Tutz
Department of Statistics
Ludwig-Maximilians-Universität
Munich, Germany

ISSN 2199-7357 ISSN 2199-7365 (electronic)
Statistics for Social and Behavioral Sciences
ISBN 978-3-031-87270-9 ISBN 978-3-031-87271-6 (eBook)
https://doi.org/10.1007/978-3-031-87271-6

This Springer imprint is published by the registered company Springer Nature Switzerland AG
The registered company address is: Gewerbestrasse 11, 6330 Cham, Switzerland

Preface

Modern Item Response Theory (IRT) provides a sophisticated framework for measuring latent traits, such as abilities, knowledge, or attitudes. It is widely used in education, psychology, healthcare, and other fields where measurement is essential. Unlike traditional testing approaches, item response theory models account for the varying difficulty levels of test items and the diverse abilities of test-takers. They are designed to investigate the relationship between latent traits and observed responses to test items with the aim to enhance the efficiency of measurement.

A main objective of the book is to provide a structured survey of diverse models that have been put forth, emphasizing both their distinctions and commonalities. It serves as an introductory guide for beginners while also serving as a resource for those seeking an overview over the plethora of available models.

The book covers foundational concepts, essential principles, and practical applications of Item Response Theory. In presenting individual models, brevity is maintained, with a focus on conciseness rather than formal complexity. The objective is to be accessible to a broad audience, emphasizing fundamental concepts and the interrelationships between models. Starred sections contain more details for the more interested reader.

Some of the topics that are covered in this book are:

- Foundations of IRT: We begin with the binary Rasch model, illustrating the fundamental principles that underpin Item Response Theory, from latent trait basics to mathematical formulations.
- Ordinal IRT Models: The key idea is to model the probability of an individual's response falling into a specific category for each test item. This modeling is based on the assumption that individuals possess a latent trait (ability or attribute) that influences their likelihood of responding in a particular category. The ordinal nature of the responses implies that the categories have a natural ordering.
- Response Models for Count Data: It is an extension of traditional Item Response Theory that is designed to handle situations where the outcome variable is a count (i.e., a non-negative integer). While conventional IRT models are typically used for binary or ordinal outcomes, IRT for count data accommodates discrete counts,

making it suitable for scenarios where the responses are, for example, the number of correct answers on a test or the frequency of certain behaviors.
- Extended versions of ordinal response models that account for response styles are included.
- Thresholds Model: The thresholds model is an overarching model that contains various common latent trait models as special cases and provides a genuine latent trait model for continuous responses.
- Tree-based Item Response Models: They include the special topic of Item Response Trees, which are a type of modeling approach that combines aspects of binary or polytomous Item Response Theory and decision tree methodologies. The approach is particularly useful when dealing with complex and multidimensional latent traits.
- Differential Item Functioning (DIF): DIF refers to a situation where an item on a test or assessment functions differently for different groups of individuals, even when those individuals have the same level of the latent trait being measured. DIF in particular occurs when subgroups of individuals with the same underlying trait level have different probabilities of responding correctly to a particular test item.
- Explanatory Item Response Models: They are a class of statistical models that extend traditional Item Response Theory by incorporating explanatory variables or covariates into the model. The primary goal is to investigate and account for the influence of additional factors on individuals' responses to test items. These models are particularly useful when researchers want to understand how external variables affect the probability of a certain response category, beyond the individual's latent trait.

The focus of the book is on modern latent trait theory models which provide measurement tools that clearly separate between person abilities and item parameters. The much older classical test theory is considered only briefly in particular to clarify the distinction between the differing approaches to measurement and investigate their relationship. The book does not give in-depth analyses of specific data sets but the application of models is illustrated by using several data sets from differing areas. They show how models can be fitted and compared.

In recent years R has become a widely used analysis tool in psychology and the social sciences. R is open source and various methods have been implemented in useful packages. All the examples in the book have been computed by using R. For most of the examples code is made accessible on Github (https://github.com/GerhardTutz/Item-Response-Models). In the book itself only snippets of code are given but the book can be read without being familiar with R coding. No introduction to R is included but there are many texts that are easily accessible. It should also be mentioned that continually new packages are implemented, which is also a reason why not much space is devoted to R code. Better programs might be around the corner.

The primary emphasis is placed on unidimensional models which assume that the latent trait being measured is unidimensional, meaning that it can be adequately represented by a single underlying construct. These models are useful when the goal is to measure a single latent trait, such as ability, knowledge, or a specific characteristic. Multi-dimensional models are considered as tools to account for specific behavioral characteristics of respondents like response styles, but also then it is assumed that the trait that is to be measured is unidimensional.

The book is aimed at applied statisticians, researchers working in psychometrics, for educators, graduate students, and anyone curious about modeling strategies that enhance the precision and validity of their measurement tools and have to select among the multitude of available modeling approaches.

Essential features that distinguish the presentation from other books on latent trait models include the following:

- The role of binary models as foundational elements of latent trait models is emphasized, as many models currently in use can be derived from these binary models. It is a structured presentation of models starting with binary models, which are at the heart of almost all IRT models.
- In particular, polychotomous models not only incorporate binary models but can also be constructed from them. This construction provides a clearer understanding of polychotomous models compared to previous representations and results in a simple taxonomy.
- Ordinal models that account for response styles are considered in a separate chapter.
- The thresholds model is presented as a general model that allows the consideration of discrete and continuous responses within a closed framework. It also serves as a foundational model for the more traditional classical test theory.
- Hierarchical tree-based latent trait models, typically not included in older textbooks, are considered.
- Latent trait models for count data are included.
- The concept of specific objectivity, also referred to as parameter separability, is examined with a distinction between theoretical and empirical parameter separability.
- Furthermore, this exploration encompasses more recently developed methods that emerged at the intersection of advanced data analysis and machine learning, such as the regularization of estimates.

Thanks to Pascal Jordan, Can Gürer, Martin Spieß, Maria Iannario, Boris Forthmann, Clemens Draxler and Ingrid Maurer for valuable comments that helped to improve the presentation.

Munich, Germany Gerhard Tutz

Contents

Chapter 1
Introduction

Modern item response theory aims at modeling the relationship between latent traits and observed responses to test items. It focuses on measuring traits that are not directly observable by using observable responses. The latent trait to be measured can be a mental ability, a proficiency, or skill but also an attitude, an emotional state, physical fitness, or other "constructs". It does not necessarily refer to persons but can also refer to properties of other entities as software when software reliability is to be investigated or to institutions concerning their efficiency in performing. Achieving reliable measurement often requires a substantial number of items. Measurement, in the context of IRT, is not an end in itself but typically serves a specific purpose, whether it be decision-making, theory testing, evaluation, or monitoring.

The establishment of the now-dominant concept of a distinct separation between the trait to be measured and the measurement instrument took time. The roots of item response theory can be traced back at least a century. Early attempts to measure abilities were based on the assumption that observed test scores can be decomposed into a true score and a random error (Spearman, 1904), leading to classical test theory (CTT). By assuming an additive model the score itself represents the measurement apart from random fluctuation. Thurstone used the cumulative normal distribution function (normal ogive) to separate intelligence from age (Thurstone, 1925), but in particular his work in psychophysics (Thurstone, 1927), including the idea of a normally distributed latent response process, was pivotal for the development of item response theory.

Clear measurement concepts, assuming that test items are located on the same scale as the construct they measure, were later formulated by Lord (1952) and Rasch (1960). The concepts of ability and difficulty attained the same status as the fundamental concepts of force and mass in physics, quantities that cannot be observed directly (see also Van der Linden (2016b)). These features distinguish item response theory from classical test theory (Lord and Novick, 1968), which employs an a priori score on the entire test by assuming an additive decomposition of an observed test score into a true score and a random error.

G. Tutz, *A Short Guide to Item Response Theory Models*, Statistics for Social and Behavioral Sciences, https://doi.org/10.1007/978-3-031-87271-6_1

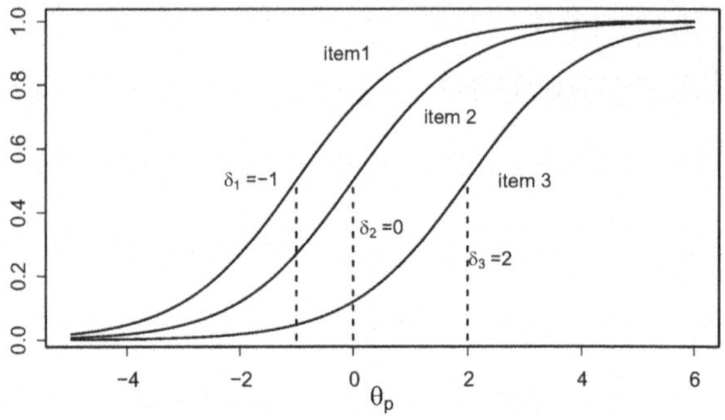

Fig. 1.1 Item characteristic curves for three items with difficulties $\delta_1 = -1, \delta_2 = 0, \delta_3 = 2$

The evolution of item response theory over the years has been traced back carefully by Van der Linden (2016b) and Thissen and Steinberg (2020). Notable contributions are found in particular in the work of Spearman (Spearman, 1904), Thurstone (Thurstone, 1925), Lord and Novick (Lord and Novick, 1968), and Rasch (Rasch, 1960, 1961a).

Main Features of Item Response Theory Models

Essential components of modern item response theory are

- The trait to be investigated (ability, skill, attitude) is latent.
- Items can be located on the same scale as the latent trait.
- The latent trait accounts for observed interrelationships among the item responses.
- The responses are random and have to be described by a probabilistic model to explain their distributions.

An important model that exemplarily shows all these properties is the binary Rasch model. It specifies how the observable response Y_{pi}, if person p tries to solve item i, is determined by properties of the person and the item. In the simplest case, the response is in two categories, 0 and 1. In achievement tests, 1 typically denotes success and 0 failure, in attitude measurement often 1 denotes agreement and 0 disagreement. The model can be given in the form:

$$P(Y_{pi} = 1|\theta_p, \delta_i) = F(\theta_p - \delta_i), \tag{1.1}$$

where θ_p is the person parameter and δ_i is the item parameter, which in achievement tests are interpreted as ability and difficulty, respectively. The response function $F(.)$ is the logistic distribution function $F(\eta) = \exp(\eta)/(1 + \exp(\eta))$, which is strictly increasing over the whole domain. The model specifies that the probability of success is a function of the difference between ability and difficulty, the larger the difference the larger the probability of success. Both parameters, the latent trait θ_p and the item parameter δ_i, are located on the same scale. It takes abilities that are much larger than the item parameter to obtain a large probability of success. Figure 1.1 shows the probability of success for three items with difficulties $\delta_1 = -1, \delta_2 = 0, \delta_3 = 2$ as a function of the person parameter θ_p. It is seen that for each item the probability of success in that item is 0.5 if the person parameter equals the item parameter and increases to 1 if the person parameter increases.

The separation between the latent trait and the measurement instrument as present in the Rasch model is central to the idea of measurement in modern item response theory. In general, the latent trait refers to a *property* or *attribute* of a person or another entity to be measured. Not the person is measured but an attribute. Various terms are in common use to describe the property of interest. Properties are also referred to as *attributes*, *latent traits*, *latent variables*, or *constructs*, the last term indicating that traits are of a hypothetical nature. In this text, we will mostly use the term latent trait, but also simply refer to ability implying that an achievement test is considered.

Assessment and Item Types

A distinction that is useful is between assessment and measurement as emphasized, for example, by Andrich and Marais (2019). Assessment refers to the controlled procedure that is used to make inferences on the latent trait. A specific trait is assessed by using a specific test, which means collecting data. The assessment procedure yields observations in the form of responses, which are typically scored responses. In the simplest case, the scores are 0 and 1 with 1 indicating better performance. Items of this type are *binary* or *dichotomous* and are scored dichotomously. Items can also have more than two levels of performance. Typically the numbers $0, 1, \dots k$ are used to represent the levels of performance with 0 denoting the weakest performance. Items of this type are called *polytomous* items and are scored polytomously. Also in questionnaires with response categories ranging from *totally disagree* to *strongly agree* the scores are typically given as integers, ranging, for example, from 0 to 6 when using a 7-point scale. In some cases, it is useful to assess performance by counts represented by $0, 1, \dots$. For example, in memory tasks, the performance is assessed by the number of remembered items. Items of this type have been used very early in the development of modern item response theory (Rasch, 1960). Other types of items take real-valued numbers, for example, response-time data or responses within a line segment, say between 0 and 100. In the former case, responses are restricted to the positive domain, in the latter to an interval.

Items are in particular characterized by their support S, that is, the range of values they can take. The item types in common used are the following:

$S = \{0, 1\}$ if responses are binary or dichotomously scored;
$S = \{0, 1, \dots, k\}$ for polytomous responses;
$S = \{0, 1, \dots, \}$ for count data;
$S = [0, \infty)$ for continuous, non-negative responses;
$S = (a, b)$ for responses within an interval;
$S = (-\infty, \infty)$ for continuous, unrestricted responses.

Measurement

While the scores represent the observations they are not the measurement. Measurement means linking the observations to the latent trait. The Rasch model (1.1) is an example how this link can be formulated by specifying a probabilistic model.

Probabilistic models are a specific way to measurement in the behavioral and social sciences, which differs from approaches typically used in the physical sciences.

The problem of measurement is an old one and various concepts of measurement have been proposed. An interesting overview has been given by Hand (1996). He distinguishes in particular between representational theory and operational theory. *Representational theory* is the one most researchers have in mind. It links an empirical relational system (ERS), which contains the relations between objects that are postulated or observed to a numerical relational system (NRS), which comprises numbers and numeric relations between them. In a very simple example, rods are compared regarding their length with the empirical relation being "is longer than" which is linked to the numerical system with the numeric relation $\geq$, which is a relation between numbers. Measurement is established by finding a mapping from the objects to the numerical system such that the relationships between objects correspond to the relationship between numbers, technically the mapping is a so-called homomorphism. Since differing numerical representation of relationships between objects is possible, for example using centimeters instead of meters, one obtains a taxonomy on the representations yielding the familiar concept of *scale types*. Already Stevens, in very early papers on mathematics, measurement, and psychophysics, distinguished between nominal, ordinal, interval, and ratio scales, a distinction with which most researchers are familiar (Stevens, 1946, 1951). In *nominal* scales, the assigned numbers only indicate that two objects to which different numbers are assigned are different from each other with regard to the attribute to be measured. In *ordinal* scales, the order of objects with respect to the attribute is represented but differences between numbers are not meaningful. In contrast, *interval* scales have a unit but an arbitrary origin. Thus differences of numbers can be interpreted. In *ratio* scales as length or weight also ratios of numbers make sense.

Representational theory relies on axioms that the empirical system has to satisfy to be adequately represented in the numerical system. In particular, axioms that relate to extensive structures and contain concatenation operations have been derived and shown to be useful in physics. In the social sciences concatenation, operations like placing rods end to end are hardly available. The role of axioms is taken by assumptions on the probabilities that have to be satisfied in order to allow for a numerical representation, an example are the derivations of the Rasch model from sets of assumptions given by Fischer (1995). Although some derivations of the Rasch model only postulate specific behavior of subjects, others assume specific properties of probabilities given the latent trait, thus postulating properties of the measurement itself, in contrast to pure representational theory, which uses axioms that describe the behavior of objects only. Representational measurement theory was outlined extensively in the three volumes of foundations of measurement Krantz et al. (1971), Suppes and Krantz (2007), Luce et al. (2007).

While relational theory aims at measuring an attribute that is present in objects or subjects *operational measurement theory* is somewhat weaker in its demands. It identifies the attribute to be measured by the measurement procedure, and measurement becomes any specified operation that yields a number (Dingle, 1950). Then the existence of the attribute plays a different role. It is closer to the concept that we construct our reality rather than detect reality. Even carefully constructed tests do to a certain degree establish the attribute and similar tests might refer to somewhat different constructed attributes. Although this seems far fetched when measuring weight or length, it is not so hard to accept that two questionnaires, seeking to tap into some attitude to a proposition but different questions, are describing slightly different phenomena (Hand, 1996). Operational theory emphasizes that we are using "constructs" when measuring latent traits. There are also alternative theories of measurement as the classical theory (Michell, 2014). It emphasizes that attributes not objects are measured and is a realist theory that assumes that pre-existing relationships are discovered by linking hypothesized attributes to observations. As outlined by Hand (1996), Rasch's concept of specific objectivity, which will be discussed later, can be seen within this framework.

Structuring Models: A Taxonomy of Models

The various types of items have led to a variety of models and several taxonomies have been proposed to bring some structure into the plethora of available models. In the presentation given here, the role of binary models as building blocks of latent trait models is emphasized. Most models in current use can be constructed from binary models by using them as models for dichotomized responses, for example, by distinguishing only between responses below or above some threshold. If a model for, say, polytomous responses holds it typically also contains binary models for subsets of response categories. Thus, the binary models can be seen from a construction point of view but also from a deconstruction point of view, with the latter aiming at identifying the binary models that are implicitly assumed to hold. It is not surprising that binary models are at the heart of model building. The latent trait is supposed to model the impact of a person's ability to perform well. Separating superior performance from weak performance means to distinguish between high and low test scores, or, more concrete, between responses above or below some threshold. Most models then can be seen as containing binary models for different thresholds but containing the same person parameter. A simple example are polytomous items with levels of performance $0, 1, \ldots, k$. It might be assumed that the ability of a person determines the response probability when splitting response categories into $0, \ldots, r$ and $r + 1, \ldots, k$ for any $r < k$. Assuming binary models with the same person parameter but different item parameters in the splits yields the graded response model, which is one of the most often used models for polytomous responses.

The various item response theory models differ in particular in the way the binary models are embedded. One can distinguish between unconditional models and conditional models. The latter assume a (binary) distinction between high and low performance in a conditional way, for example, by distinguishing between categories r and $r + 1$ *given* the response is in categories $r, r + 1$. Dichotomizations of this type are typically used in ordinal polytomous models and tree-based models. In particular, tree-based models make conditioning a construction principle.

Figure 1.2 shows the models that can be seen as derived from binary models. They include models for polytomously scored items, for count data and continuous responses. The models in the upper part, which are shown by rectangles around the

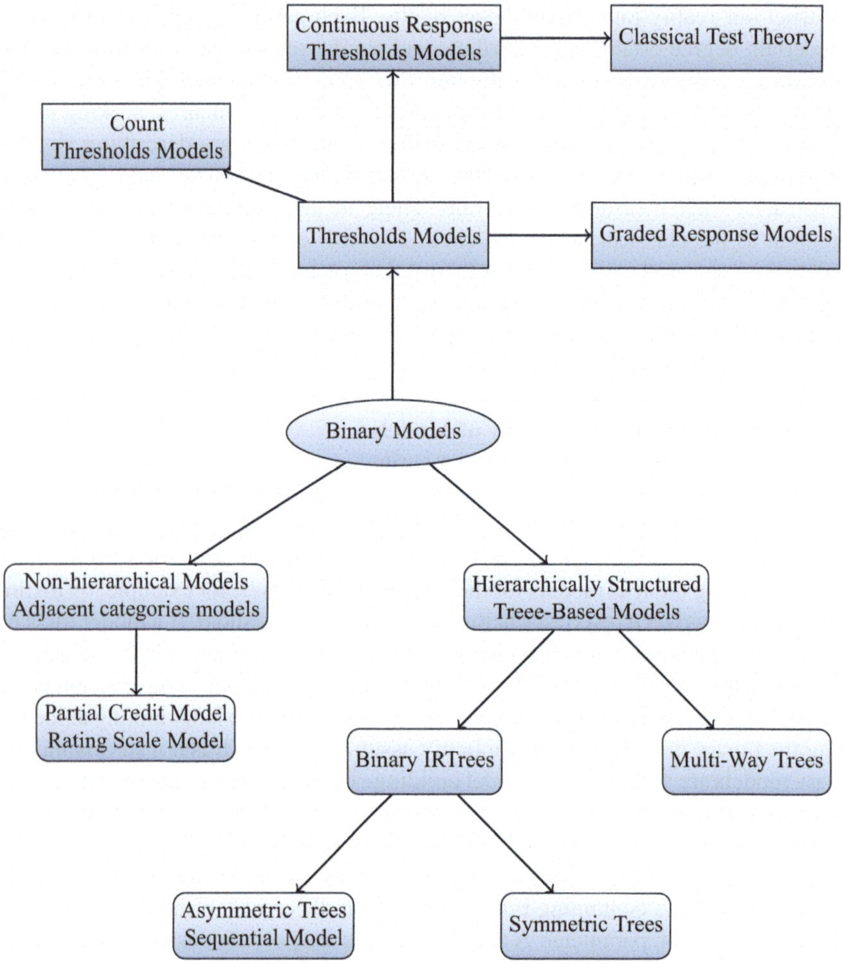

Fig. 1.2 Models for various types of responses generated by binary models

text, are unconditional models. They use the splitting of responses below and above some value. The models in the lower part (rounded corners) are based on conditioning. In particular, all tree-based models are conditional models. The structure also includes the classical ordinal models. The graded response model is a model based on splitting of categories and is a submodel of the thresholds model. The adjacent categories model uses conditioning (given the response is in adjacent categories). The sequential model is a specific tree-based model that assumes an underlying sequential process.

The strength of identifying the binary models that are embedded in general models is that it helps in the interpretation of person and item parameters. The well-established interpretation of the parameters in binary models can be used to obtain a

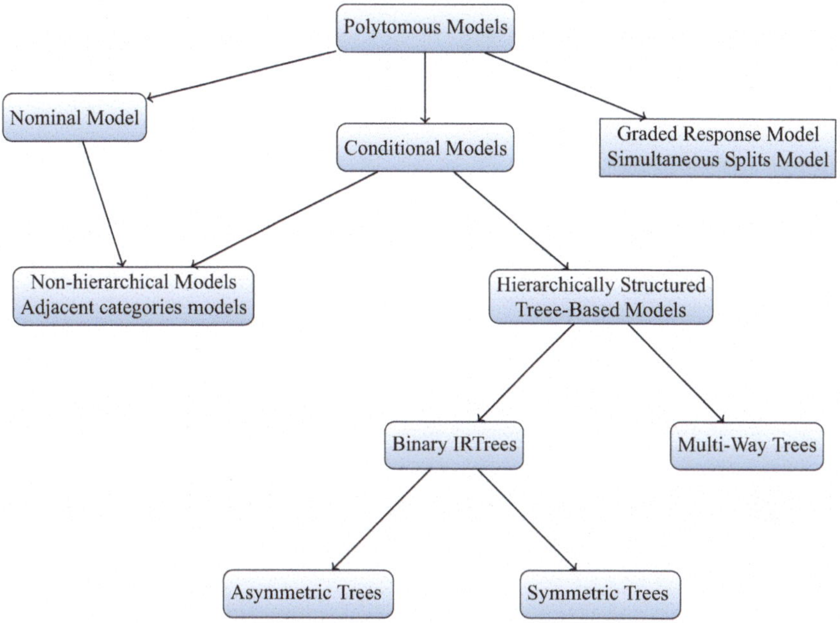

Fig. 1.3 Hierarchy of polytomous models

clearer understanding of the parameters, in particular the item parameters, in more complex models. This works for the multitude of models for which the embedded binary models have the simple (or a slightly more extended parametric) form of the binary model given above. Further models have been proposed in the literature, for example, for count data, in which the embedded binary models have a quite different, more complicated form. Models of this type will also be considered but in an all-comprising taxonomy they would have to be considered a separate class of models.

Also if one restricts consideration to polytomous models with responses $0, \ldots, k$ the binary models that are embedded in the models help to derive a simple taxonomy of models. The structuring again distinguishes between models that use binary models in a conditional way, for example, by modeling responses in r given responses in adjacent categories $r, r+1$, or unconditionally (rounded corners). Figure 1.3 shows the resulting hierarchy of models, which includes the nominal model. The latter has not been included in Fig. 1.2 since no simple binary model is embedded. The taxonomy for polytomously scored items will be discussed in detail in Chaps. 4 and 9.

Chapter 2
The Binary Rasch Model

The binary Rasch model is one of the cornerstones of modern item response theory and has been extended to a whole family of models. It has been designed as a probabilistic measurement model that clearly separates person and item parameters allowing to infer on item parameters without reference to person parameters. The Rasch model and the competing normal-ogive model were the first latent trait models that have been formulated in a clear-cut way. In this chapter, the Rasch model is considered, alternative binary models will be considered separately in the following chapter.

2.1 The Model

Let persons or test-takers $1, \ldots, P$ be confronted with a set of similar tasks or items $1, \ldots, I$ to be mastered. It is only distinguished between success and failure. The response Y_{pi} of person p on item i takes values from $\{0, 1\}$, where (in proficiency tests) "1" indicates success and "0" indicates failure.

The crucial assumption in latent trait models as the Rasch model is that items can be located on the same scale as the latent trait. In the case of the Rasch model, it is assumed that the probability of success has the following form:

Binary Rasch Model

$$P(Y_{pi} = 1|\theta_p, \delta_i) = \frac{\exp(\theta_p - \delta_i)}{1 + \exp(\theta_p - \delta_i)} = \frac{e^{\theta_p - \delta_i}}{1 + e^{\theta_p - \delta_i}}, \tag{2.1}$$

where θ_p is the ability of person p and δ_i is the difficulty of item i.

G. Tutz, *A Short Guide to Item Response Theory Models*, Statistics for Social and Behavioral Sciences, https://doi.org/10.1007/978-3-031-87271-6_2

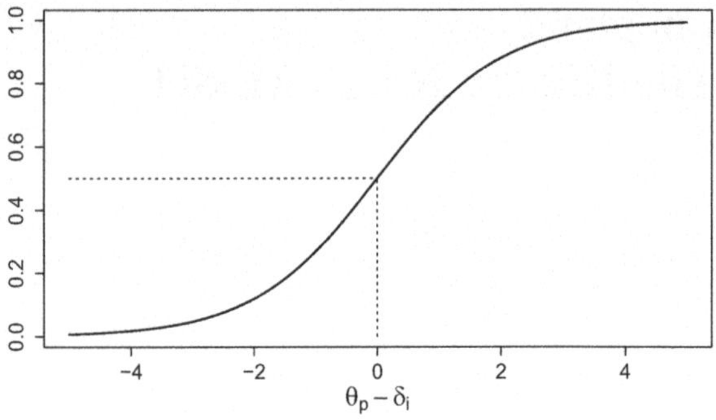

Fig. 2.1 Logistic function $F(\eta) = \exp(\eta)/(1 + \exp(\eta))$

The model specifies the probability of success as a simple function of the difference between the ability and difficulty $\theta_p - \delta_i$. The function that transforms the differences into probabilities is the logistic function $F(\eta) = \exp(\eta)/(1 + \exp(\eta))$. By using this function the model can also be given in the form:

$$P(Y_{pi} = 1|\theta_p, \delta_i) = F(\theta_p - \delta_i). \tag{2.2}$$

Figure 2.1 shows the logistic function, which is a sigmoidal, symmetric function centered at zero with $F(0) = 0.5$. It is seen that the probability of success is close to one if the ability θ_p is much larger than the item difficulty δ_i since the difference $\theta_p - \delta_i$ becomes large. If θ_p is much smaller than δ_i the difference is negative and the probability of success becomes very small.

Item Characteristic Curves and Invariant Item Ordering
An important characterization of a binary model is obtained by considering the *item characteristic curve* (ICC) or *item response function* (IRF). This function specifies the probability of success as the level of the ability increases. It can be considered the graph of the probability of success depending on θ_p. For the Rasch model, the item characteristic curve of item i is given by

$$ICC_i(\theta_p) = P(Y_{pi} = 1|\theta_p, \delta_i) = F(\theta_p - \delta_i). \tag{2.3}$$

Each item has its own item characteristic function with θ_p as argument and δ_i considered as fixed. Figure 2.2 shows the item characteristic curve for an item with difficulty $\delta_i = 1$. It is seen that the probability of success increases strictly with the level of the ability. If the ability equals the difficulty, $\theta_p = \delta_i = 1$, the probability of success is 0.5. For larger values of θ_p, the probability is higher, one obtains, for example, 0.73 if $\theta_p = 3$. For very large values of θ_p, it approaches one, thus an examinee with a very large ability is supposed to answer correctly.

Figure 2.3 shows the item characteristic curves for three items with difficulties $\delta_1 = -1, \delta_2 = 0, \delta_3 = 2$. Item 1 is the easiest and item 3 the hardest to solve. For example, a person with $\theta_p = 0$ has a probability larger than 0.5 to solve item 1, probability 0.5 to solve item 2 but a rather small probability of solving item 3. It

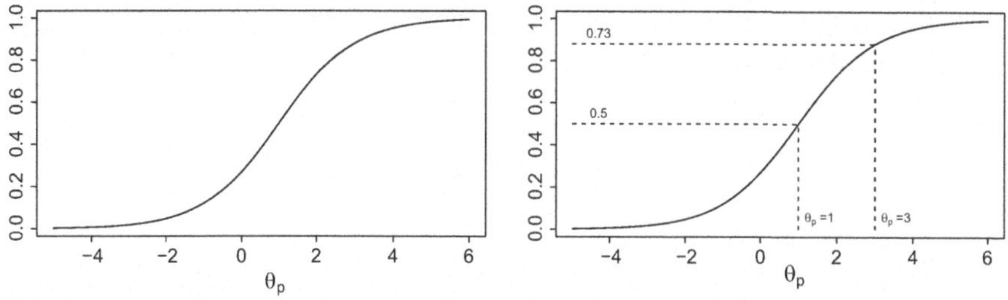

Fig. 2.2 Item characteristic curve for an item with difficulty $\delta_i = 1$

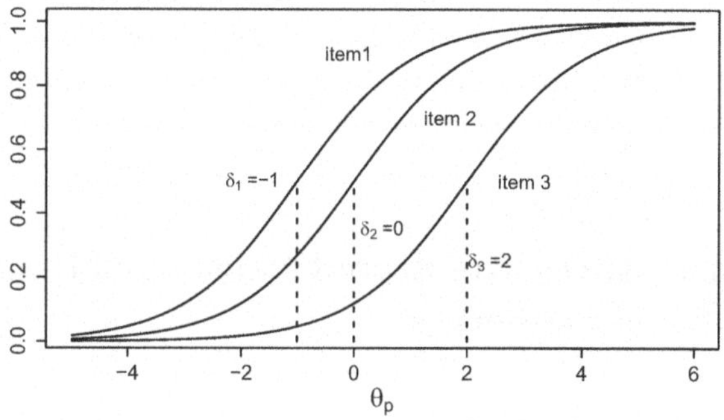

Fig. 2.3 Item characteristic curves for three items with difficulties $\delta_1 = -1, \delta_2 = 0, \delta_3 = 2$

is noteworthy that item characteristic curves for different items never cross, which means for any person the probabilities of success for items reflect the order of item difficulties. More formally, if items are ordered such that

$$\delta_{[1]} \leq \cdots \leq \delta_{[I]},$$

representing increasing ability, the probabilities of success for the items are ordered by

$$P(Y_{p[1]} = 1|\theta_p, \delta_{[1]}) \geq \cdots \geq P(Y_{p[I]} = 1|\theta_p, \delta_{[I]}) \quad \text{for any } \theta_p. \tag{2.4}$$

If (2.4) holds the ordering is an *invariant item ordering* (IIO), which is often considered a desirable property of latent trait models. It means that one has the same ordering of the items by difficulty across different levels of the trait or ability. Thus, if item 1 is harder than item 2 it is harder to solve for able persons as well as for less able persons.

Local Stochastic Independence

The item response curves show how the ability determines the probability of success for a single item. It says nothing about the distribution of the responses for the other items. For a complete specification of the model, the common distribution of item responses has to be specified. The assumption that is typically made is that responses are independent given item and person parameters. This assumption of *local stochastic independence* means that the responses collected in the matrix $\boldsymbol{Y} = (Y_{pi})$ are determined by

$$P(\boldsymbol{Y} = \boldsymbol{y}|\boldsymbol{\theta}, \boldsymbol{\delta}) = \prod_{p=1}^{P} \prod_{i=1}^{I} P(Y_{pi} = y_{pi}|\theta_p, \delta_i),$$

where $\boldsymbol{\theta}^T = (\theta_1, \dots, \theta_P)$, $\boldsymbol{\delta}^T = (\delta_1, \dots, \delta_I)$. It means that the systematic covariation between item responses is explained by the variation of the latent trait. However, if persons are drawn at random from a population the *marginal* responses will not be independent but positively correlated since persons with a higher ability tend to have higher success probability in all items while persons with a lower ability tend to have lower success probabilities in all items. Local stochastic independence means that responses are *conditionally* independent given item and person parameters.

2.2 Alternative Representations and Uniqueness of Parameters

A model representation that specifies the probability of success and failure simultaneously is given by

$$P(Y_{pi} = y_{pi}|\theta_p, \delta_i) = \frac{\exp(y_{pi}(\theta_p - \delta_i))}{1 + \exp(\theta_p - \delta_i)}, \quad y_{pi} \in \{0, 1\}, \tag{2.5}$$

which yields

$$P(Y_{pi} = 1|\theta_p, \delta_i) = \frac{\exp(\theta_p - \delta_i)}{1 + \exp(\theta_p - \delta_i)} \text{ and } P(Y_{pi} = 0|\theta_p, \delta_i) = \frac{1}{1 + \exp(\theta_p - \delta_i)}.$$

Representations that follow from the specific choice of the logistic function as response function link the parameter to odds, which are common tools in binary regression. Using the response probabilities of success and failure, one obtains the model representations

$$\frac{P(Y_{pi} = 1|\theta_p, \delta_i)}{P(Y_{pi} = 0|\theta_p, \delta_i)} = \exp(\theta_p - \delta_i) = \frac{\exp(\theta_p)}{\exp(\delta_i)} \tag{2.6}$$

and

$$\log\left(\frac{P(Y_{pi} = 1|\theta_p, \delta_i)}{P(Y_{pi} = 0|\theta_p, \delta_i)}\right) = \theta_p - \delta_i. \tag{2.7}$$

The first representation shows that the *odds* $P(Y_{pi} = 1)/P(Y_{pi} = 0)$, which compare the probability of success and failure as a proportion, are given by

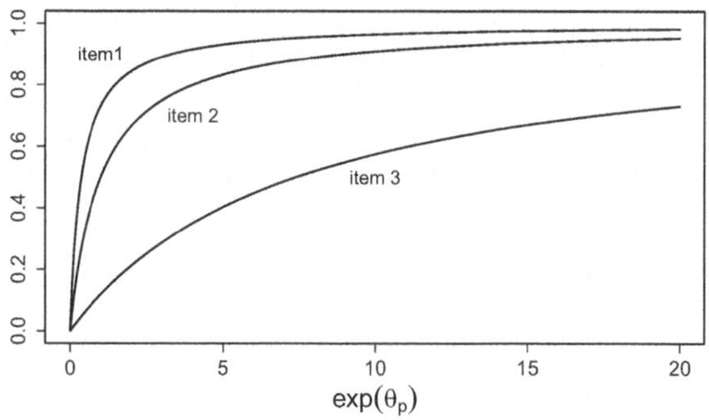

Fig. 2.4 Item characteristic curves for three items for the same difficulties as in Fig. 2.3 but with transformed parameters $\tilde{\theta}_p = \exp(\theta_p)$, $\tilde{\delta}_i = \exp(\delta_i)$

$\exp(\theta_p)/\exp(\delta_i)$, which is a proportion of person and item parameters. From the second representation (2.7), it is seen that the *log-odds* $\log(P(Y_{pi} = 1)/P(Y_{pi} = 0))$ are equivalent to the difference of ability and item difficulty. Thus, in the Rasch model, parameters are linked in a simple way to odds and log-odds.

The Rasch model has also been given in alternative parameterizations. By defining $\theta_p^* = \exp(\theta_p)$ and $\delta_i^* = \exp(\delta_i)$ one obtains the representation

$$P(Y_{pi} = 1|\theta_p^*, \delta_i^*) = \frac{\theta_p^*}{\theta_p^* + \delta_i^*}, \tag{2.8}$$

where $\theta_p^*, \delta_i^* \geq 0$. It is the same model but abilities and difficulties are given on a different scale. The corresponding odds have the form:

$$\frac{P(Y_{pi} = 1|\theta_p, \delta_i)}{P(Y_{pi} = 0|\theta_p, \delta_i)} = \frac{\theta_p^*}{\delta_i^*}.$$

It shows that for fixed item parameter the odds are directly proportional to $\theta_p^* = \exp(\theta_p)$. If $\theta_p^* = 0$ ($\theta_p \to -\infty$) the odds and the probability of success are zero, if $\theta_p^* \to \infty$ ($\theta_p \to \infty$) the odds increase indefinitely and the probability of success becomes 1.

Figure 2.4 shows the item response curves for the transformed ability parameter $\theta_p^* = \exp(\theta_p)$. Though the same difficulties as in Fig. 2.3 have been used item response curves have a quite different shape since the parameters have been transformed. However, again item response curves do not cross and invariant item ordering holds also in alternative parameterizations.

Whatever representation is used, the parameters of the model are not identifiable. In the (θ_p, δ_i) parameterization used in representation (2.5), one can replace the parameters θ_p, δ_i by parameters $\bar{\theta}_p = \theta_p + c$ and $\bar{\delta}_i = \delta_i + c$ using any constant c without changing the probabilities. Since $\bar{\theta}_p - \bar{\delta}_i = \theta_p - \delta_i$ the model with parameters θ_p, δ_i is indistinguishable from the model with parameters $\bar{\theta}_p, \bar{\delta}_i$. In other words, the complete scale can be shifted to the left or right, yielding identical probabilities. To obtain a unique parameterization some constraint is needed. One can set one of the parameters to a fixed value, for example, $\theta_1 = 0$ or $\delta_1 = 0$. Symmetric constraints which also ensure identifiability are $\sum_p \theta_p = 0$ or $\sum_i \delta_i = 0$. They set the mean of either person or item parameters to zero. Also in the multiplicative parameterization (θ_p^*, δ_i^*) in (2.8) parameters are not identified since one can replace the parameters θ_p^*, δ_i^* by parameters $\bar{\theta}_p = c\theta_p^*$ and $\bar{\delta}_i = c\delta_i^*$, $c > 0$. Identifiability is obtained by using constraints as $\theta_1^* = 1$ or $\delta_1^* = 1$. An alternative symmetric constraint is $\prod_p \theta_p^* = c$ for some positive value c. Constraints make the parameters unique but do not reflect the scale level that is used.

In order to see which scale level is used when the Rasch model is considered a measurement model it is helpful to present the model in the more general form:

$$P(Y_{pi} = 1|\theta_p, \delta_i) = \frac{\exp(a(\theta_p - \delta_i) + b)}{1 + \exp(a(\theta_p - \delta_i) + b)},$$

which can be considered a "family of Rasch models" (Fischer, 1995). In the model, the parameters $a > 0$ and b can be chosen arbitrarily. For any choice of a, b, the transformed parameters $\theta_p^* = a\theta_p + b_1$, $\delta_i^* = a\delta_i + b_2$ yield the same probabilities since $a(\theta_p - \delta_i) + b = \theta_p^* - \delta_i^*$, where $b = b_1 - b_2$. Thus, the θ_p- and the δ_i-scales are determined up to positive linear transformations with the same multiplicative constant a. The choice of a corresponds to the unit of measurement while b refers to the choice of the origin and the parameters θ_p and δ_i have interval scale properties with a common unit of measurement, for more details, see Fischer (1995).

2.3 Information Function

Concepts that are useful when using likelihood-based estimation methods are information and the information function. Item characteristic curves are steepest at the value $\theta_p = \delta_i$, that is, if the ability equals the item difficulty (Fig. 2.3). If θ_p is far away from δ_i increasing ability changes the response probability only marginally. Thus, for a fixed item the information on the ability gathered if a person tries to solve the item will be largest if the person's ability is close to the difficulty. If it is much larger or smaller than the difficulty the gain in information will be much weaker.

The information function is a technical tool that defines what information means in the context of binary models. For item i item information (or Fisher information) can be defined as the expectation

$$I_i(\theta_p) = \mathrm{E}(-\frac{\partial^2 l_i(Y_{pi}; \theta_p)}{\partial\theta_p \partial\theta_p}),$$

where $l_i(Y_{pi}; \theta_p)$ is the log-likelihood for item i and observation Y_{pi}. It is a function of the person parameter. For binary $Y_{pi} \in \{0, 1\}$, the log-likelihood is given by

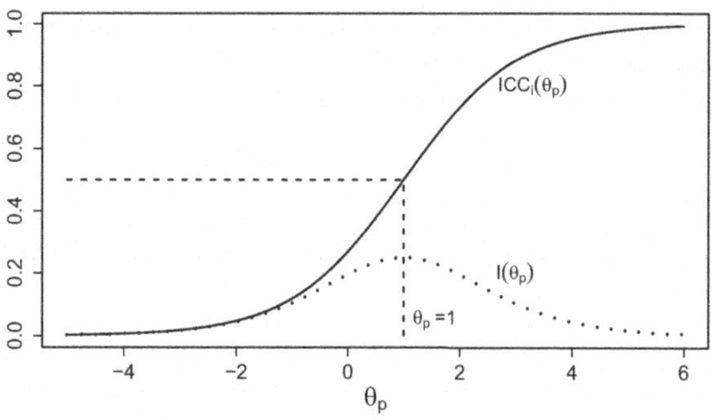

Fig. 2.5 Item characteristic curve (dashed) and information function (dotted) for an item with difficulty $\delta_i = 1$

$l_i(Y_{pi}; \theta_p) = Y_{pi} \log(\pi_i(\theta_p)) + (1 - Y_{pi})(\log(1 - \pi_i(\theta_p)))$, where $\pi_i(\theta_p) = P(Y_{pi} = 1|\theta_p, \delta_i)$. For the Rasch model, one obtains the simple form

$$I_i(\theta_p) = \pi_i(\theta_p)(1 - \pi_i(\theta_p)),$$

which is the variance of a binary variable with success probability $\pi_i(\theta_p)$. It is well known that the variance is largest if the outcome is most uncertain, that is, if the probability is 0.5. This is the case if the ability equals the difficulty. Figure 2.5 shows the information function together with the item characteristic curve for an item with difficulty $\delta_i = 1$. It is seen that the maximum of the information function is at $\theta_p = 1$ and tends to zero for very large and small values of θ_p. For the Rasch model, it is also equivalent to the derivative of the item characteristic function:

$$I_i(\theta_p) = \frac{\partial}{\partial \theta_p} P(Y_{pi} = 1|\theta_p, \delta_i).$$

The information is important in the estimation of parameters since under regularity conditions the asymptotic standard error of maximum likelihood estimates $\hat{\theta}_p$ is inversely proportional to the square root of the information, $\text{var}(\hat{\theta}_p) \propto 1/I_i(\theta_p)$. This is the technical reason why it can be called information since larger information means higher accuracy of estimates. The information in more general binary models has been considered, for example, by Lord (1980) and Magis (2013).

2.4 Separability of Parameters and Model Derivations

The response function in the Rasch model is the logistic function. It is a very specific choice that has the consequence that the model has several useful properties. One of them is the existence of sufficient statistics, which can be interpreted as separability of parameters or specific objectivity of the model. In the following, concepts of specific objectivity are considered. They are linked to an alternative approach to latent trait modeling, namely, to postulate that a latent trait model should have specific properties and derive a model from these properties. We will briefly consider which postulates lead to the logistic response function used in the Rasch model.

2.4.1 Separability of Parameters and Specific Objectivity

A property that has been considered as important when considering latent trait models as measurement instruments is the so-called *specific objectivity* (Rasch, 1966, 1977). With reference to stimuli rather than items Rasch (1961b) formulated the requirements for comparing individuals and comparing stimuli by stating:

"The comparison between two stimuli should be independent of which particular individuals were instrumental for the comparison, and it should also be independent of which other stimuli within the considered class were or might also have been compared."

"Symmetrically, a comparison between two individuals should be independent of which particular stimuli within the class considered were instrumental for the comparison, and it should also be independent of which other individuals were also compared, on the same or on some other occasion (Rasch, 1961b), p.331."

The Rasch model allows for comparisons of item difficulties (or person abilities) that are independent of the selection of persons (or items). Rasch used the term *specific objectivity* for measurements that allow for the comparison of subjects or objects without reference to the instrument (within a given well-defined frame of reference), and considered it as general scientific principle to obtain generalizable measurements. The property has also been referred to as *parameter separability* or *invariance of comparisons*.

It can be given in several forms, but the main issue is that the two agent sets, items and persons, can be separated such that comparisons within one set do not depend on the choice of agents from the other set that is "used" to compare them. The property can be seen as making the measurement independent of the measurement instrument.

More concrete, let us again consider the odds, which for a probability of success π are generally defined by $\gamma = \pi/(1-\pi)$. For the Rasch model, the odds are given by

$$\gamma_{pi} = \frac{P(Y_{pi} = 1|\theta_p, \delta_i)}{P(Y_{pi} = 0|\theta_p, \delta_i)} = \exp(\theta_p - \delta_i). \tag{2.9}$$

It follows that the odds ratio for two persons p_1, p_2 is given by

$$\frac{\gamma_{p_1 i}}{\gamma_{p_2 i}} = \frac{\exp(\theta_{p_1})}{\exp(\theta_{p_2})} = \frac{\theta^*_{p_1}}{\theta^*_{p_2}}, \tag{2.10}$$

which does not depend on the item. Thus, comparisons of persons can be carried out independently of the item involved.

In a similar way, the odds ratio for two items i_1, i_2 is given by

$$\frac{\gamma_{pi_1}}{\gamma_{pi_2}} = \frac{\exp(-\delta_{i_1})}{\exp(-\delta_{i_2})} = \frac{\delta^*_{i_2}}{\delta^*_{i_1}}, \tag{2.11}$$

which does not depend on the person. Thus, comparisons of items can be carried out independently of the person involved.

The comparison of items not depending on the persons makes measurement independent of the population, which is similar to measurement in physics, where the measurement is supposed to not depend on the measurement instrument. However, there is one difference. In probabilistic measurement models, as the Rasch model, the parameters have to be estimated, they are not known, and measurement relies on estimation. Therefore, parameter separability and specific objectivity have sometimes been described in terms of estimation concepts.

One concept that works in Rasch models is conditional estimation. It can be shown that the conditional distribution of responses of a person p given s, the number of correctly solved items by the person has the form:

$$P(Y_{p1} = y_{p1}, \ldots, Y_{pI} = y_{pI} | \sum_{i=1}^{I} Y_{pi} = s) = e^{-\sum_{i=1}^{I} y_{pi}\delta_i} / \gamma_s(\boldsymbol{\delta}),$$

where $\gamma_s(\boldsymbol{\delta})$ is a specific function of the item parameters $\boldsymbol{\delta}^T = (\delta_1, \ldots, \delta_I)$ (see Sect. 2.5). It means that the distribution of responses given the number of correctly solved items does not depend on the person parameter but is a function of the item parameters only. In technical terms, the sum score $S_p = \sum_i Y_{pi}$ is a *sufficient statistic* for the person parameters. In general, statistics are defined as sufficient statistics for a set of parameters if the conditional distribution of observations given the sufficient statistic does not depend on the parameter set. It means that all the information on the parameters available from observations is contained in the sufficient statistics. Conditioning on the sufficient statistic yields a (conditional) log-likelihood that does not contain the parameter (Sect. 2.5). Thus, person parameters are not involved when the (conditional) log-likelihood is maximized. That seems to make the estimation of item parameters independent of the sample of persons. In a technical sense this is true, the estimation equations do not depend on the item parameters. However, the accuracy of estimates may still depend on the persons that have been drawn. It makes a difference if persons are from a small range of abilities or from a wide range of abilities, if they are in the range of the item parameters or far away from them (see Example 2.2). That means estimation of parameters should not be considered as independent of the person parameters although many authors assume just that.

The problem with comparisons of item (person) parameters being considered as independent of person (item) parameters is that typically it is not distinguished between separability of parameters as a property of the probabilistic measurement model and *empirical separability* as an estimation problem. In the Rasch model, *separability of parameters in the measurement model* is captured by Eqs. (2.10) and

(2.11). Empirical separability refers to the existence of estimation procedures that do not involve the other set of parameters. In the case of the Rasch model, this is obtained by conditional log-likelihood estimation, which allows to estimate item parameters without using person parameters. In that sense, parameters are separated, but the existence of estimation equations that do not involve person parameters should not be misunderstood, it does not mean that estimation does not depend on the sample of persons. The accuracy of estimates depends on the choice of the sample (see also Example 2.2).

The Rasch model is the only binary model known to allow for sufficient statistics and therefore conditional estimation of parameters with the estimation equations not depending on the other set of parameters. This has led many researchers to consider the model as the only one that has the property of specific objectivity. This holds only if specific objectivity and separation of parameters is considered as being equivalent to the existence of suitable sufficient statistics. Separability of parameters as a property of the measurement model holds for the much larger class of homogeneous monotone latent trait model to be considered in Chap. 3.

2.4.2 Derivations and Model Motivations*

One of the most striking features of the Rasch model is the existence of sufficient statistics, that is, that the information on a set of parameters contained in a sample is captured by a specific combination of observations. This feature also plays an important role if the model is considered from a measurement perspective. The measurement approach to item response theory starts with a theoretical construct and a set of measurement desiderata. A measurement instrument is constructed that matches the desiderata or postulates (Glas, 2007). Several sets of postulates yield the Rasch model, one is the following.

Let the item characteristic curves for all items be functions of one person parameter and assume that the following set of postulates holds:

(M) Item characteristic curves are continuous, strictly monotone increasing functions for all items and persons.

(L) The lower limits of the item characteristic functions are determined by $\lim_{\theta_p \to -\infty} ICC_i(\theta_p) = 0$ and upper limits by $\lim_{\theta_p \to \infty} ICC_i(\theta_p) = 1$.

(SI) Local stochastic independence holds for all responses.

(S) The raw score statistic $S_p = \sum_{i=1}^{I} Y_{pi}$ is a sufficient statistic for the person parameter.

If these assumptions hold the item response model is the binary Rasch model (see Fischer (1995), p 16 Theorem 2.1). The first two assumptions are rather mild restrictions that seem sensible in all item response models (without guessing). Local stochastic independence is an assumption that postulates that the correlation of responses is due to the variation of abilities in the population. The key assumption is that the sum of correctly solved items is a sufficient statistic, that is, the sum score contains all the information in the data about the person parameter. Although the assumption (S) can be replaced by the postulate of stochastically consistent ordering (Roskam and Jansen, 1984) it can be shown to be equivalent to the existence of sufficient statistics (Fischer, 1995).

Thus, sufficiency of sums is one of the defining properties of the Rasch model, which means

The sum of correctly solved items $S_p = Y_{p+} = \sum_{i=1}^{I} Y_{pi}$ is a sufficient statistic for θ_p. It contains all the information on the person parameter.

The sum of persons that solved the ith item, $Y_{+i} = \sum_{p=1}^{I} Y_{pi}$, is a sufficient statistic for δ_i. It contains all the information on the item parameter.

An advantage of the sufficiency of sums is that it suffices to know the sum of correctly solved items if one wants to learn about the ability of a person.

It is instructive that the Rasch model (and other binary models) can also be derived from the assumption of the separability of parameters. The basic construction principle was given by Rasch (1960), Rasch (1961a), Rasch (1977), a rigorous mathematical derivation with concise proofs was given by Fischer (1995). It is assumed that persons are characterized by one parameter ξ_p and items by one parameter β_i. The response Y_{pi} if person p is confronted with item i is characterized by a response parameter ρ_{pi} which is a function of these parameters, that is, $\rho_{pi} = G(\xi_p, \beta_i)$. In probabilistic models, the response parameter can, for example, be thought of being the probability of solving the item, which is a function of the parameters ξ_p, β_i. It is further assumed that the function $G(.,.)$ is continuous; the function $G(., \beta)$ is an increasing bijective function for any β; and the function $G(\xi, .)$ is a decreasing bijective function for any ξ. The essential assumption is that parameters are separable, which means that it is assumed that a function $C(.,.)$, the so-called *comparator*, exists such that

$$C(G(\xi, \beta), G(\tilde{\xi}, \beta)) \text{ is a function of } \xi, \tilde{\xi} \text{ only.}$$

Since $C(G(\xi, \beta), G(\tilde{\xi}, \beta))$ is independent of β it can be represented as a function $V(\xi, \tilde{\xi})$, that is,

$$C(G(\xi, \beta), G(\tilde{\xi}, \beta)) = V(\xi, \tilde{\xi}). \tag{2.12}$$

In addition, the comparator $C(.,.)$ is assumed to be continuous, strictly monotonically increasing in the first component and decreasing in the second component. Equation (2.12) is the essence of specific objectivity. If two persons responding to one item can be compared by a comparator function yielding a function that does only depend on the person parameters the response system allows for the separation of parameters.

An important result is that given these assumptions hold the comparator and the reaction parameter can be represented by

$$C(\rho_{p_1 i}, \rho_{p_2 i}) = c(\theta_{p_1} - \theta_{p_2}), \quad \rho_{pi} = g(\theta_p - \delta_i),$$

where $c(.)$, $g(.)$ are continuous bijective increasing functions and the parameters θ_p, δ_i are obtained from continuous bijective increasing transformation functions $t(.)$, $d(.)$ that yield the parameters $\theta_p = t(\xi_p)$, $\delta_i = d(\beta_i)$ (see Lemma 2.4, Fischer (1995)). As Fischer (1995) remarks the result is very general, without any specificity for item response theory, it also applies to the laws in physics, for example, the dependence of force on mass and acceleration.

If the reaction parameter is the probability of success, $\rho_{pi} = \pi_{pi}$, a probability model results that contains the difference between person and item parameters,

$$\pi_{pi} = g(\theta_p - \delta_i), \tag{2.13}$$

with $g(.)$ denoting an increasing function that is not necessarily the logistic function. It means that other models that contain the difference of parameters, and therefore show *latent additivity*, could be derived from separability of parameters.

The Rasch model with its specific logistic response function results if one lets the comparator be a function of likelihoods. Considering the sample space for two subjects and one item, $A = \{(1, 1), (1, 0), (0, 1), (0, 0)\}$, and subsets $C \subset B \subseteq A$ one obtains the Rasch model (apart from linear transformations of parameters and a fixed multiplicative constant) if the comparator is a function of the unconditional likelihood $l(B)$ or a conditional likelihood $l(C|B)$ (see Theorem 2.5, Fischer (1995)). Using the likelihood means that it is an *empirical* separation of parameters, not a theoretical one as (2.13), which is based on probabilities. It links the Rasch model to likelihoods and therefore to likelihood estimation concepts. The distinction between empirical and theoretical concepts of parameter separation will also be considered in Chap. 3 and Sect. ??.

2.5 Estimation

When fitting binary item response models an additional assumption concerning the dependence of the observations is needed. The most common assumption is *local stochastic independence*, that is, it is assumed that $Y_{p1}, \ldots, Y_{pI}$ are independent given the parameters:

$$P(Y_{p1} = y_{p1}, \ldots, Y_{pI} = y_{pI}|\theta_p, \delta_1, \ldots, \delta_I) = P(Y_{p1} = y_{p|1}|\theta_p, \delta_1) \ldots P(Y_{pI} = y_{pI}|\theta_p, \delta_I).$$

In addition, it is assumed that the vectors $\boldsymbol{Y}_1, \ldots, \boldsymbol{Y}_P$, where $\boldsymbol{Y}_p^T = (Y_{p1}, \ldots, Y_{pI})$, are independent.

If all persons try to solve the items one obtains the data matrix given in Table 2.1.

Since $Y_{pi} \in \{0, 1\}$ the marginal sums $Y_{p+} = \sum_{i=1}^{I} Y_{pi}$, $Y_{+i} = \sum_{p=1}^{P} Y_{pi}$ have a clear meaning:

Y_{p+} represents the number of items solved by person P and
Y_{+i} represents the number of persons that solved item i.

Table 2.1 Data matrix for binary responses

		Item 1	2	...	I	
Person	1	y_{11}	y_{12}	...	y_{1I}	y_{1+}
	2	y_{21}	$\ddots$		$\vdots$	$\vdots$
	$\vdots$	$\vdots$		$\ddots$	$\vdots$	
	P	y_{P1}	...		y_{PI}	y_{P+}
		y_{+1}	...		y_{+I}	

2.5.1 *Joint Maximum Likelihood Estimation*

Under the assumption of local stochastic independence the likelihood that aims at estimating person and item parameters simultaneously has the form

$$L = \prod_{p=1}^{P}\prod_{i=1}^{I}\frac{\exp(y_{pi}(\theta_p - \delta_i)}{1+\exp(\theta_p - \delta_i)} = \frac{\exp(\sum_{p=1}^{P} y_{p+}\theta_p - \sum_{i=1}^{I} y_{+i}\delta_i)}{\prod_{p=1}^{P}\prod_{i=1}^{I}(1+\exp(\theta_p - \delta_i))},$$

where y_{p+} represents the number of items solved by person p and y_{+i} represents the number of persons that solved item i.

It is straightforward to maximize the corresponding log-likelihood

$$l = \log(L) = \sum_{p=1}^{P} y_{p+}\theta_p - \sum_{i=1}^{I} y_{+i}\delta_i - \log(\prod_{p=1}^{P}\prod_{i=1}^{I}(1+\exp(\theta_p - \delta_i))),$$

and compute the derivatives needed to obtain variance estimates. However, joint maximum likelihood (JML) estimation has some severe drawbacks. With $I, P \to \infty$ the number of parameters is growing, which makes it hard to find an asymptotic distribution. In particular, it yields biased estimates if I is much smaller than P, with the bias being of order $I/(I-1)$ (Andersen, 1977; Haberman, 1977). The estimator is inconsistent for $P \to \infty$, I fixed, but consistency holds for $I, P \to \infty$, $P/I \to \infty$ (Haberman, 1977).

2.5.2 *Conditional Maximum Likelihood Estimation*

Conditional maximum likelihood estimation (CML) exploits the existence of sufficient statistics in the Rasch model. Sufficient statistics contain the total information on parameters, by conditioning on them the dependence on these parameters is canceled.

Let us consider the response distribution for a specific person p with sum score $s = y_{p+} = \sum_{i=1}^{I} y_{pi}$ (number of solved items). The distribution is given by

$$P(Y_{p1} = y_{p1}, \ldots, Y_{pI} = y_{pI}) = P(Y_{p1} = y_{p1}) \ldots P(Y_{pI} = y_{pI})$$
$$= \frac{e^{y_{p1}(\theta_p - \delta_1)}}{1 + e^{\theta_p - \delta_1}} \cdots \frac{e^{y_{pI}(\theta_p - \delta_I)}}{1 + e^{\theta_p - \delta_I}} = \frac{e^{s\theta_p} e^{-\sum_{i=1}^{I} y_{pi}\delta_i}}{\prod_{i=1}^{I}(1 + e^{\theta_p - \delta_i})} = \frac{e^{s\theta_p} e^{-\sum_{i=1}^{I} y_{pi}\delta_i}}{\gamma_{0p}},$$

where $\gamma_{0p} = \prod_{i=1}^{I}(1 + \exp(\delta_p - \delta_i))$. For the distribution of the sum score, one obtains

$$P(\sum_{i=1}^{I} Y_{pi} = s) = \sum_{y_{p1}+\cdots+y_{pI}=s} P(Y_{p1} = y_{p1}) \ldots P(Y_{pI} = y_{pI})$$
$$= \frac{e^{s\theta_p} \sum_{y_{p1}+\cdots+y_{pI}=s} e^{-\sum_{i=1}^{I} y_{pi}\delta_i}}{\prod_{i=1}^{I}(1 + \exp(\theta_p - \delta_i))} = \frac{e^{s\theta_p} \sum_{y_{p1}+\cdots+y_{pI}=s} e^{-\sum_{i=1}^{I} y_{pi}\delta_i}}{\gamma_{0p}}$$
$$= \frac{e^{s\theta_p}\gamma_s(\boldsymbol{\delta})}{\gamma_{0p}},$$

where the functions

$$\gamma_s(\boldsymbol{\delta}) = \sum_{a_i \in \{0,1\}, \sum_i a_i = s} e^{-\sum_i a_i \delta_i}$$

are the so-called elementary symmetric functions of order s, depending on $\boldsymbol{\delta}^T = (\delta_1, \ldots, \delta_I)$ only. The consequence is that the conditional distribution given the sum score is given by

$$P(Y_{p1} = y_{p1}, \ldots, Y_{pI} = y_{pI} | \sum_{i=1}^{I} Y_{pi} = s) = \frac{e^{-\sum_{i=1}^{I} y_{pi}\delta_i}}{\gamma_s(\boldsymbol{\delta})},$$

which is a function of the item p only and does not depend on the person parameter θ_p. In the Rasch model, the sum score $s = y_{p+}$ is a sufficient statistic for θ_p, by conditioning on the sum score the dependence of the response probabilities on θ_p vanishes.

The corresponding conditional likelihood has the form:

$$L_c = \prod_{p=1}^{P} P(Y_{p1} = y_{p1}, \ldots, Y_{pI} = y_{pI} | \sum_{i=1}^{I} Y_{pi} = y_{p+}) = \prod_{p=1}^{P} \frac{e^{-\sum_{i=1}^{I} y_{pi}\delta_i}}{\gamma_{y_{p+}}(\boldsymbol{\delta})}. \tag{2.14}$$

Maximization yields estimates of the item parameters $\hat{\delta}_1, \ldots, \hat{\delta}_I$. However, for items that have been solved by all persons or by none of them no estimates are obtained.

Necessary and sufficient conditions for the existence of estimates were given by Fischer (1981). The estimator is consistent for $P \to \infty$, I fixed (Pfanzagl, 1994). The asymptotic distribution and approximate estimation errors were given by Andersen (1977). For the loss of information in CML, see Eggen (2000), Eggen and Verhelst (2006).

2.5.3 Marginal Maximum Likelihood Estimation

Marginal maximum likelihood (MML) estimates of item parameters are obtained by assuming a fixed distribution of person parameters. The corresponding marginal likelihood has the form

$$L_m = \prod_{p=1}^{P} \int \prod_{i=1}^{I} P(Y_{pi} = 1|\theta_p, \delta_i)^{y_{pi}} P(Y_{pi} = 0|\theta_p, \delta_i)^{1-y_{pi}} f(\theta_p) d\theta_p,$$

where $f(\theta_p)$ is the density of the distribution assumed for the person parameters. Typically, one assumes a normal distribution with mean zero and variance σ_θ^2, which can be abbreviated by $f_{0,\sigma_\theta}(.)$. Many program packages allow only normal distributions although the distribution of the person parameters has not to be normal, see, for example, Andersen and Madsen (1977), Thissen (1982).

Semi-parametric or nonparametric estimation approaches that have been proposed use more flexible distributions of item parameters. One approach replaces the integral over the mixing distribution by a finite sum,

$$L_n = \prod_{p=1}^{P} \sum_{g=1}^{G} \prod_{i=1}^{I} P(Y_{pi} = 1|\theta_g, \delta_i)^{y_{pi}} P(Y_{pi} = 0|\theta_g, \delta_i)^{1-y_{pi}} \pi_g,$$

where $g = 1, \ldots, G$ are latent groups with locations θ_g and $\pi_1 \ldots, \pi_G$ are the mixture probabilities with $\sum_g \pi_g = 1$ (De Leeuw and Verhelst, 1986; Follmann, 1988). For more details, including nonparametric approaches to estimate person parameters, see Molenaar (1995).

2.5.4 Conditional Pairwise Maximum Likelihood Estimation

As conditional ML estimation, conditional pairwise ML estimation exploits that the sum score is a sufficient statistic for θ_p. However, only pairs of items are considered. For two items i, j and any person p, the sum score (number of solved items) is given by $s = Y_{pi} + Y_{pj}$. From the general result (2.14), one obtains $P(Y_{pi} = y_{pi}, Y_{pj} = y_{pj}|s) = e^{-y_{pi}\delta_i - y_{pj}\delta_j}/\gamma_s(\boldsymbol{\delta})$. In the special case $s = 1$, the probabilities are given by

$$P(Y_{pi} = 1, Y_{pj} = 0|Y_{pi} + Y_{pj} = 1) = P(Y_{pi} = 1|Y_{pi} + Y_{pj} = 1) = \frac{e^{-\delta_i}}{e^{-\delta_i} + e^{-\delta_j}},$$

which do not depend on θ_p. Thus, for pairs of observations Y_{pi}, Y_{pj} the (conditional) probability of $Y_{pi} = 1$ given $Y_{pi} + Y_{pj} = 1$ does not depend on the selected person. Therefore, for a random sample of persons from a population in which the model holds, one considers only those persons for which exactly one of a pair of items has been solved, that is, $Y_{pi} + Y_{pj} = 1$. Let N_i denote number of persons that solved item i and exactly one of the two items. Then the relative frequency $R_i = N_i/P$, where P is the number of items for which $Y_{pi} + Y_{pj} = 1$ holds, is an unbiased estimator of $P(Y_{pi} = 1|Y_{pi} + Y_{pj} = 1)$. The ratio of two item parameters, which are given by

$$\frac{P(Y_{pi} = 1|Y_{pi} + Y_{pj} = 1)}{P(Y_{pj} = 1|Y_{pi} + Y_{pj} = 1)} = \frac{e^{-\delta_i}}{e^{-\delta_j}},$$

can be estimated by $R_i/R_j = N_i/N_j$. Thus the ratio of any two item parameters can be estimated independently of the sample that was drawn. The estimate $e^{-\hat{\delta}_i} e^{-\hat{\delta}_j} N_i/N_j$ can also be derived from maximizing the conditional likelihood for the pairs of items given $Y_{pi} + Y_{pj} = 1$. It is important that the number of observations is so large that for any pairs of items the number of observations with $Y_{pi_1} + Y_{pi_2} = 1$ is not too small. For more on pairwise estimates, see Zwinderman (1995), von Davier (2016).

2.5.5 *Estimation of Person Parameter*

Joint maximum likelihood estimation yields estimates of person and item parameters, but has the disadvantage that estimates are not consistent for I fixed. In addition, all item parameters typically change if persons are added. Therefore, usually item parameters are estimated using a large sample of test-takers using conditional estimation methods. Then, item parameters are considered as known and the person parameter is estimated from the responses of an individual to the items.

One estimator that is constructed in this way is the maximum likelihood estimator of person parameters. For a fixed person, the corresponding likelihood is

$$L(\theta_p) = \prod_{i=1}^{I} \frac{\exp(y_{pi}(\theta_p - \delta_i))}{1 + \exp(\theta_p - \delta_i)} = \frac{\exp(y_{p+}\theta_p - \sum_{i=1}^{I} y_{pi}\delta_i)}{\prod_{i=1}^{I}(1 + \exp(\theta_p - \delta_i))},$$

yielding the log-likelihood

$$l = \log(L) = y_{p+}\theta_p - \sum_{i=1}^{I} y_{pi}\delta_i - \sum_{i=1}^{I} \log(1 + \exp(\theta_p - \delta_i))$$

and Fisher information

$$I(\theta_p) = \mathrm{E}(-\frac{\partial^2 l(\theta_p)}{\partial\theta_p \partial\theta_p}) = \sum_{i=1}^{I} \frac{\exp(\theta_p - \delta_i)}{(1 + \exp(\theta_p - \delta_i))^2}.$$

For large I, the variance of the estimator can be approximated by the inverse of the Fisher information. With $\hat{\theta}_p$ denoting the maximum likelihood estimator one can use $1/I(\hat{\theta}_p)$ as an approximation to $\mathrm{var}(\hat{\theta}_p)$. It can also be shown that $\hat{\theta}_p$ asymptotically follows a normal distribution. More concise, $(\hat{\theta}_p - \theta_p)\sqrt{I(\theta_p)}$ follows asymptotically a standardized normal distribution $N(0, 1)$, which may be used to compute confidence intervals, for details see, for example, Hoijtink and Boomsma (1995). For $y_{p+} = 0$ and $y_{p+} = I$, the maximum likelihood estimator does not exist since it tends to $-\infty$ or ∞, respectively. However, the probability of zero and perfect raw scores tends to zero for $I \to \infty$.

An alternative are Bayesian estimators that use a fixed density of person parameters $f(\theta_p)$, which is assumed to be known or estimated from a sample. Then, given item responses $\boldsymbol{y}_p^T = (y_{p1}, \ldots, y_{pI})$ and item parameters $\boldsymbol{\delta}^T = (\delta_1, \ldots, \delta_I)$ the posterior of θ_p is given by

$$f(\theta_p|\boldsymbol{y}_p, \boldsymbol{\delta}) = \frac{L(\theta_p)f(\theta_p)}{\int L(\theta_p)f(\theta_p)d\theta_p} = \frac{\exp(y_{p+}\theta_p)b(\theta_p)f(\theta_p)}{\int \exp(y_{p+}\theta_p)b(\theta_p)f(\theta_p)d\theta_p},$$

where $b(\theta_p) = (\prod_{i=1}^{I}(1 + \exp(\theta_p - \delta_i)))^{-1}$. Maximization of $f(\theta_p|\boldsymbol{y}_p, \boldsymbol{\delta})$ yields the *Bayes modal estimate*. It has the advantage that estimates exist also for persons with zero and perfect raw scores. However, as in marginal estimation of item parameters a specific prior distribution of person parameters has to be chosen. A specific prior is the “uninformative” Jeffrey’s prior, which for the Rasch model is given by $f(\theta_p) = \sqrt{I(\theta_p)}$. Warm (1989) showed that it has smaller bias than the maximum likelihood estimator.

Instead of the mode one can also compute the posterior mean

$$\hat{\theta}_m = \mathrm{E}(\theta_p|\boldsymbol{y}_p, \boldsymbol{\delta}) = \int \theta_p f(\theta_p|\boldsymbol{y}_p, \boldsymbol{\delta})d\theta_p,$$

yielding the *expected a posteriori estimator*.

Conditional pairwise estimators of person parameters have been considered by Draxler et al. (2016). A useful overview on person parameter estimates including finite sample properties has been given by Hoijtink and Boomsma (1995).

Example 2.1 FIMS study

An FIMS dataset study contains responses on 14 mathematics items and 2 covariates (gender, 1: male, 2: female; country, 1: Australia, 2: Japan), and is available from the TAM package. We use a subset of 300 persons with binary responses (data.fims.Aus.Jpn.scored). Table 2.2 shows the estimates of item parameters together with the sum scores (number of persons that solved a specific item). The third and fourth columns show the conditional ML estimates of difficulty parameters $\hat{\delta}_i$ and the corresponding easiness parameters given by $-\hat{\delta}_i$. The last column shows the marginal estimates of difficulty parameters, which are very similar. In all estimates the restriction that estimates sum to one has been used. It is seen that item 3 is the easiest item to solve. It has the largest sum score, the lowest difficulty parameter and the largest easiness parameter.

Models can be easily fitted by using R packages. The conditional estimates given in Table 2.2 are obtained by using the package *eRm* (Mair et al., 2016).

```
load("datAust300")
datscored<-datAust300

library("eRm") #conditional likelihood estimation

claustRM<- RM(datscored,  se = TRUE, sum0 = TRUE)
summary(claustRM)

cbind(Sumscore=colSums(datscored),
   Difficulty=-round(claustRM$betapar,digits=3),
      Easiness=round(claustRM$betapar,digits=3))
```

The output contains the sum scores, the difficulty, and easiness parameters. Marginal estimates are obtained by using the package *ltm*:

```
library("ltm")

pcmltm<-gpcm(datscored, constraint = "rasch")
pcmltm

## rescaling such that sum of item parameters is zero
matr<-pcmltm$coefficients[[1]]
for (i in 2:14)matr<-rbind(matr,pcmltm$coefficients[[i]])
meanitems<-colSums(matr)[1]/14
matr[,1]<-matr[,1]-meanitems

round(matr,digits=3)
```

Table 2.2 Estimates of item parameters for FIMS study

	Conditional			Marginal
	Sum score	Difficulty	Easiness	Difficulty
1	226	−1.616	1.616	−1.651
2	218	−1.457	1.457	−1.490
3	249	−2.146	2.146	−2.189
4	170	−0.631	0.631	−0.652
5	24	2.428	−2.428	2.467
6	228	−1.658	1.658	−1.694
7	102	0.444	−0.444	0.456
8	113	0.262	−0.262	0.267
9	54	1.402	−1.402	1.442
10	184	−0.857	0.857	−0.882
11	39	1.828	−1.828	1.874
12	58	1.305	−1.305	1.343
13	32	2.076	−2.076	2.122
14	214	−1.380	1.380	−1.413

The output contains the difficulty parameters given in Table 2.2, scaled such that the sum over difficulty parameters is zero. □

For a detailed description of the R package *eRm*, see Mair et al. (2016). Packages and their applications are also discussed in the comprehensive book by Mair (2018). Baker et al. (2017) provide detailed R code examples for binary response models. An introduction to the Rasch model, with numerous examples in R, is given by Debelak et al. (2022).

The following example illustrates that the independence of item parameters on person parameters, which is often highlighted as the big advantage of the Rasch model, does not hold in a strict sense if independence refers to estimation, which means parameter separability understood as empirical separability.

Example 2.2 Independence of parameter estimates
In Rasch models, item parameters are often considered to not depend on persons when maximizing the conditional likelihood given person sums since the conditional likelihood does not contain person parameters. However, the choice of persons has an impact on the accuracy of estimates. It matters if persons are drawn from a population covering a wide range of abilities or from a population that contains only persons with high (or low) abilities.

Table 2.3 illustrates that effect. Person parameters have been drawn from several distributions, the standard normal distribution ($\theta_p \sim N(0, 1)$), which covers a wide range of person parameters, the χ^2-distribution with one degree of freedom ($\theta_p = \tilde{\theta}_p^2$ with $\tilde{\theta}_p \sim N(0, 1)$), which means that all person parameters are positive, the non-central χ^2-distribution with non-centrality parameter 1 ($\theta_p = \tilde{\theta}_p^2$ with $\tilde{\theta}_p \sim N(1, 1)$),

Table 2.3 Average deviations of estimates of item parameters from true values for several estimation methods

		Conditional	Pairwise conditional
P = 80	Standard normal	0.257	0.290
	Chi-squared	0.258	0.289
	Nonc chi-squared, $N(1, 1)^2$	0.303	0.346
	Nonc chi-squared, $N(1.5, 1)^2$	0.369	0.452
P=100	Standard normal	0.231	0.252
	Chi-squared	0.240	0.265
	Nonc chi-squared, $N(1, 1)^2$	0.253	0.291
	Nonc chi-squared, $N(1.5, 1)^2$	0.311	0.365
P=200	Standard normal	0.154	0.170
	Chi-squared	0.166	0.180
	Nonc chi-squared, $N(1, 1)^2$	0.182	0.200
	Nonc chi-squared, $N(1.5, 1)^2$	0.213	0.231

and the non-central χ^2-distribution with non-centrality parameter 1.5 ($\theta_p = \tilde{\theta}_p^2$ with $\tilde{\theta}_p \sim$ N(1.5, 1)). In particular, the latter scenarios mean that only rather large person parameters are in the sample. The item parameters of the $I = 6$ items were $0, -1.5, -1, 0.5, 1.2, 1.5$. As measure of accuracy we use the absolute deviation $|\delta_i - \hat{\delta}_i|$ averaged across all items. Results are given for several sample sizes. It is seen that all estimates perform poorer if the means of person parameters become larger. In particular, the pairwise conditional estimator suffers strongly if the mean of the distribution of person parameters is large. Tutz (2024a) proposed an alternative estimator that uses the separability of parameters but does not rely on sufficient statistics. It typically outperforms the pairwise conditional estimator but not the conditional estimator. □

2.6 Testing the Rasch Model

Glas and Verhelst (1995) gave a careful overview and brought some systematics into the plethora of tests that have been proposed for the Rasch model. A useful and general survey on tools for assessing the fit of item response theory model is found in Swaminathan et al. (2006). In the following, only some tests and diagnostic tools that are in common use are considered.

2.6.1 Diagnostic Tools

Andersen's Likelihood Ratio Test

Andersen's likelihood ratio test is an overall goodness-of-fit test that uses that item parameters can be consistently estimated in any sample of persons provided the Rasch model holds in the corresponding population. It compares the estimates obtained in subgroups of subjects, which should only vary due to random fluctuations. Subgroups can be formed by external variables or on the basis of scores.

Let the persons $\bar{P} = \{1, \ldots, P\}$ be divided into disjunct subsets $\bar{P}_1, \ldots, \bar{P}_g$ such that $\bar{P} = \bar{P}_1 \cup \cdots \cup \bar{P}_g$.

Let $\boldsymbol{\delta}_j$ denote the estimated item parameters when the model is fitted by using the data Y_{pi}, $p \in \bar{P}_j, i \in \{1, \ldots, I\}$, $L_c(\boldsymbol{\delta}_j)$ denotes the corresponding conditional likelihood and $l_c(\boldsymbol{\delta}_j) = \log L_c(\boldsymbol{\delta}_j)$ the log-likelihood. $L_c(\boldsymbol{\delta})$ denotes the likelihood when all data are used and $l_c(\boldsymbol{\delta}) = \log L_c(\boldsymbol{\delta})$ the corresponding log-likelihood. Then,

$$\text{LR} = -2 \log \frac{L_c(\boldsymbol{\delta})}{\prod_{j=1}^{g} L_c(\boldsymbol{\delta}_j)} = 2(\sum_{j=1}^{g} l_c(\boldsymbol{\delta}_j) - l_c(\boldsymbol{\delta}))$$

is asymptotically χ-squared distributed with the degrees of freedom given by the number of parameters estimated in the subgroups of persons minus the number of parameters estimated in the total dataset (Andersen, 1973). Large values of LR (larger than the $(1 - \alpha)$-quantile of the χ-squared distribution for some significance level α) indicate that the parameters vary over subgroups.

Graphical Model Checks

The estimates obtained in subgroups of subjects can be visualized and compared in a graphical model check. Since apart from random fluctuations estimates of item parameters should be the same, the value pairs should be close to the bisector. Strong differences between estimates in subgroups refer to a systematic violation of the model assumptions. They might indicate that tests work differently in subpopulations, for example, if subgroups are defined by gender or that the underlying trait is not unidimensional. Of course, the results depend on the subgroups that are investigated. Subgroups that do not rely on external variables are in particular groups that split the population into subjects by ability level, for example, groups that perform below or above average, the split criterion being below or above the median or mean of raw scores.

Even if the model holds estimated item parameters cannot be expected to be the same in both groups. As a measure for the random fluctuation that can be expected given the model holds one uses two-dimensional confidence regions, typically with confidence level 0.95. If these confidence levels do not cross the bisector the corresponding items are considered as contributing to the violation of model assumptions.

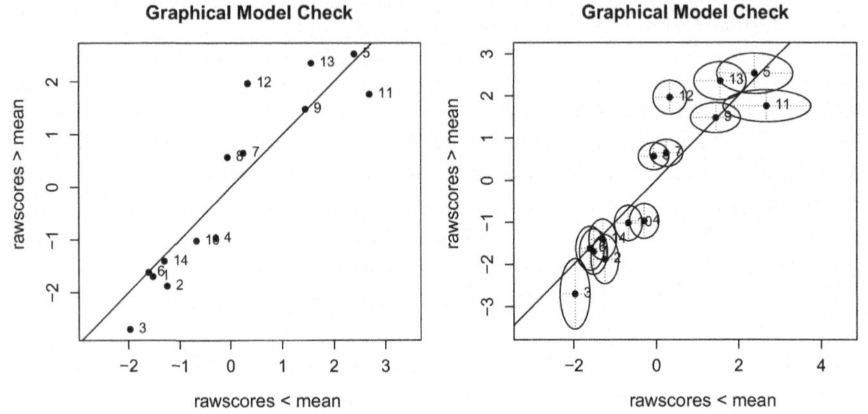

Fig. 2.6 Graphical model check for FIMS data, split criterion is the mean of raw scores, left: without confidence intervals, right: with confidence intervals

Example 2.3 FIMS study

We again use the FIMS dataset study which contains responses on 14 mathematics items used in Example 2.1. When subgroups are defined by the mean (below or above) Andersen LR test yields 57.345 on 13 df, which means the Rasch model cannot be expected to be a good representation of the data. Figure 2.6 shows the corresponding graphical model checks. It is seen that in particular items 8, 12, and 13 can be suspected of contributing to the bad fit of the model.

The likelihood ratio test and the plots are again obtained by using the package *eRm*.

```
claust<- PCM(datscored,  se = TRUE, sum0 = TRUE)  #model fit
lraust<-LRtest(claust, splitcr = "mean", se = TRUE) ## lr test
lraust

par(cex.lab=1,cex.axis=1.0,cex.main=1,6,pch=16,cex=1.6)

# plots
plotGOF(lraust, beta.subset = "all", main = "Graphical Model Check",
        xlab="rawscores < mean", ylab="rawscores > mean",
        tlab = "number",  type = "p", pos = 4, conf = NULL,
        ctrline = NULL,
        smooline = NULL, asp = 1, set_par = TRUE,  reset_par = TRUE)

plotGOF(lraust, beta.subset = "all", main = "Graphical Model Check",
        xlab="rawscores < mean", ylab="rawscores > mean",
        tlab = "number",  type = "p", pos = 4,
        conf = list(gamma=0.95,col=1), ctrline = NULL,
        smooline = NULL, asp = 1,  set_par = TRUE,
        reset_par = TRUE)
```

Example 2.4 Fears data
As an example of attitude measurement we consider data from the German Longitudinal Election Study (GLES), which is a long-term study of the German electoral process (Rattinger et al., 2014). The data we are using originate from the pre-election survey for the German federal election in 2017 and are concerned with political fears. The participants were asked: "How afraid are you due to the ..." (1) refugee crisis? (2) global climate change? (3) international terrorism? (4) globalization? (5) use of nuclear energy? Here we use binary responses, the original Likert scales from 1 (not afraid at all) to 7 (very afraid) were dichtomized by using 1 if responses were above or equal 5. When subgroups are defined by the score mean (below or above) Andersen's LR test yields 10.004 on 4 df (p-value: 0.04), which indicates that the fit of the Rasch model is not too bad. Figure 2.7 (first row) shows the corresponding graphical model checks. It is seen that only for item 1 the confidence regions do not cross the bisector. When subgroups are defined by gender Andersen's LR test yields 6.445 on 4 df (p-value: 0.168), which does not signal model violations. Figure 2.7 (second row) shows the corresponding graphical model checks. □

Comparing Items Across Subgroups
Andersen's test is based on subgroups of subjects. The same grouping can be used to examine if single items have different values in the subgroups. Let the persons be divided into just two subgroups $\bar{P}_1$, $\bar{P}_2$. Then one can test the hypothesis

$$\mathrm{H}_{0i} : \delta_i^{(1)} = \delta_i^{(2)},$$

where $\delta_i^{(j)}$ is the item parameter of item i in subgroup j by using a Wald test given by

$$\mathrm{W}_i = \frac{(\hat{\delta}_i^{(1)} - \hat{\delta}_i^{(2)})^2}{(\hat{\sigma}_i^{(1)})^2 + (\hat{\sigma}_i^{(2)})^2},$$

where $\hat{\delta}_i^{(j)}$ is the conditional ML estimate in the jth subgroup and $\hat{\sigma}_i^{(j)}$ the corresponding estimated standard error obtained from the inverse of the Fisher matrix. Under mild regularity conditions W_i is asymptotically χ-squared distributed with one degree of freedom. Parameters are not considered to be equal in subgroups if W_i exceeds the $(1 - \alpha)$-quantile of the χ-squared distribution. The resulting estimates can be represented in scatter plots. The hypothesis H_{0i} could also be tested using a likelihood ratio test but then one had to fit the models under the restriction that H_{0i} holds (i.e., that item parameters are equal across groups), which is more demanding than using the Wald test.

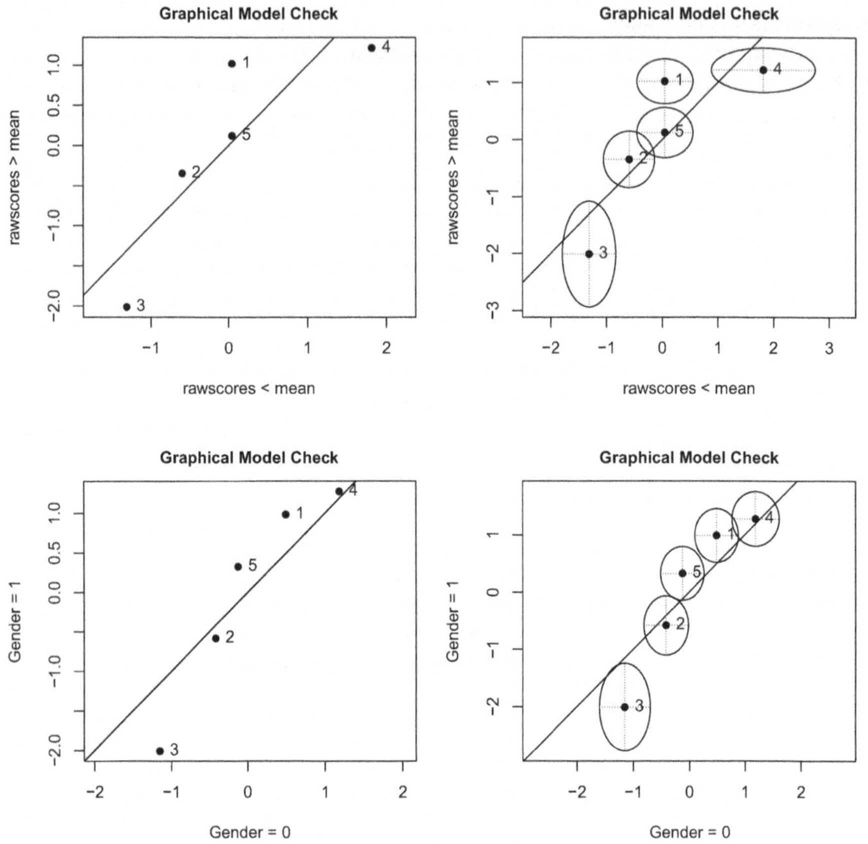

Fig. 2.7 Graphical model check for fears data, first row: subgroups defined by the mean (below or above), second row: subgroups defined by gender

Examining Subsets of Items: Martin-Löf Test

The Martin-Löf test is a likelihood ratio test for the hypothesis that the Rasch model holds for the whole set of items against the hypothesis that this model holds for two disjoint subsets of items. It examines the homogeneity of subscales based on conditional maximum likelihood estimation.

Let items be divided into two groups of items $\bar{I}_1$ and $\bar{I}_2$ that contain I_1 and I_2 items, respectively. The sum scores (number of solved items) in the two subsets form a pattern $\boldsymbol{r}^T = (r_1, r_2)$, where r_1 is the number of solved items in item group 1, r_2 the number in item group 2. The total sum of solved item is $r = r_1 + r_2$.

In the test, the occurrence of all possible patterns $\boldsymbol{r}^T = (r_1, r_2)$, $r_1 = 1, \ldots, I_1$, $r_2 = 1, \ldots, I_2$ is used. Let $n_{\boldsymbol{r}}$ denote the number of persons that attained score pattern $\boldsymbol{r}$ and n_r the number of persons that attained the score r in the full test. Then the test can be given in the form:

$$\text{LR}_{ML} = 2(\sum_r n_r \log(\frac{n_r}{P}) - \sum_r n_r \log(\frac{n_r}{P}) - \log L_C + \log L_C^{(1)} + \log L_C^{(2)}),$$

where L_C is the conditional likelihood for the total model and $\log L_C^{(j)}$ the conditional likelihood for model with items from group j. The statistic is asymptotically χ-squared distributed with degrees of freedom $I_1 I_2 - 1$.

Example 2.5 FIMS study
In Example 2.3, it was shown that for the FIMS data in particular items 8, 12, 13 can be suspected of contributing to the bad fit of the model. The Martin-Löf test can be used to investigate if the item set 8, 12, 13 and the items set 1, 2, 3, 4, 5, 6, 7, 9, 10, 11, 14 measure the same latent trait. The result was 102.473 on 32 df, which means that the hypothesis is clearly rejected. □

2.6.2 Item Selection

The Martin-Löf test is a useful tool to investigate if the Rasch model holds for the whole set of items. It investigates subscales in specific disjoint item sets. It is useful if item sets that are suspected of referring to different subscales are already given. It is less useful when the objective is to select items that share a common trait from a given set of items. The main problem is that there is a very large number of possible disjoint subsets of items. Considering subsets that contain only one item, which would reduce the number of subsets, fails since conditional maximum likelihood needs larger item sets. In addition, the tests have to assume that the model holds within the subsets, which often will not be the case. Also extended versions as considered by Christensen et al. (2002) for polytomously scored items suffer from these problems.

Anderson's test and graphical checks can be used as diagnostic tools to investigate which items do not fit well. However, the choice of the subsets crucially determines the diagnostic results. Even when comparing item sets obtained by splitting by the median or the mean different results tend to be found. Although useful when investigating specific dependencies, for example, on gender, the method is less convincing as a general item selection tool since results depend on the chosen subgroups and by using many alternate subgroups one can almost always find items that violate the assumption of the Rasch model.

In the following, a simple way of selecting items that share a common trait is described. It can be used with or without an anchor item, that is, an item that is known to measure the trait of interest. If used without an anchor item it selects all items that share one common unspecified trait.

Let parameters be estimated by using the marginal estimation approach. Then, the estimated standard deviation of the mixing distribution, σ_θ, can be used to investigate unidimensionality by exploiting that marginal estimates tend to yield stable estimates of the variance of the mixing distribution if the Rasch model holds but yield small values if items that do not share the latent trait are included. Thus, the (estimated)

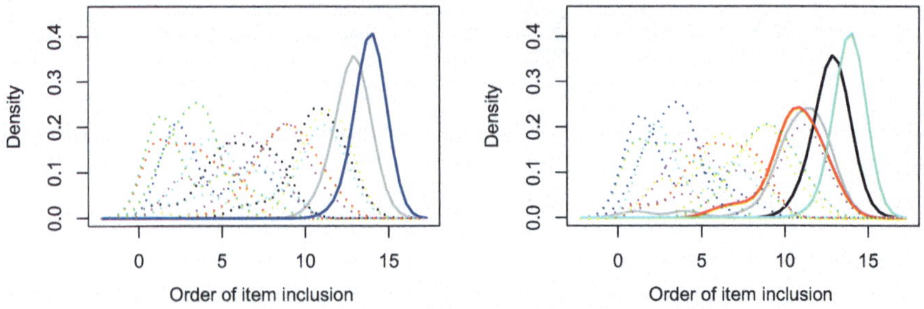

Fig. 2.8 Order distribution of item inclusion for FIMS data, left: items 8, 12 shown by drawn lines, all others dotted, right: items 7, 8, 9, 12 shown by drawn lines, all others dotted

standard deviation of the mixing distribution might be used as an indicator of the homogeneity of items, that is, their compatibility with a Rasch model. The basic algorithm includes in the first step that pair of items that has the largest standard deviation. In the following steps, the selected item set is enlarged by that item that yields the largest standard deviation if included. The algorithm is run M times using randomly selected subsets of persons. For each item and each run, it is computed at which step it has been included into the item set. One obtains for each item a distribution of the order of inclusion, for details, see Tutz (2023).

Figure 2.8 shows the distributions of the order of item inclusion for the FIMS data. It is seen that two items, item 8 and 12 (shown by drawn lines), are always included very late, all other items (dotted lines) show far less extreme distributions (left picture). As a misfit indicator, one can use the proportion of runs in which an item is included in the last quarter of inclusion steps. If the misfit indicator is large (close to 1) the item is always included very late and it is doubtful that it shares the same trait as the other items. As is seen from Table 2.4, the misfit values for items 8 and 12 are 1, they are always included very late. There might be more items that might not be compatible with the Rasch model, also items 7 and 9 have increased misfit values (Table 2.4). Figure 2.8 (right picture) shows the distributions of the order of item inclusion with items 7, 8, 9, and 12 shown by drawn lines and all other items shown by dotted lines. It is seen that the order distributions of items 7 and 9 are less extreme than for items 8 and 12 but a tendency to large order values is seen.

Table 2.4 Misfit score values for FIMS data

Item	1	2	3	4	5	5	7
Misfit	0	0	0	0	0	0.1	0.63
Mean std	0.38	0.15	0.16	0.27	0.33	0.56	0.74
Item	8	9	10	11	12	13	14
Misfit	1	0.57	0.17	0	1	0.53	0
Mean std	0.91	0.74	0.61	0.16	0.99	0.64	0.36

In Table 2.4 also standardized mean values of the order are included, which show a similar pattern as the misfit values. Standardization is again such that values are between zero and 1 and large values close to 1 indicate misfit, for more details, see Tutz (2023).

2.7 R Packages

Conditional fitting of Rasch models and ML estimation of the person parameters, as well as Andersen's LR tests, item-specific Wald tests, and Martin-Loef tests are provided by the R package *eRm* (Mair and Hatzinger, 2007b, a).

Marginal maximum likelihood and joint maximum likelihood estimates can be obtained by using R package *TAM* (Robitzsch et al., 2024) and R package *ltm* (Rizopoulos, 2006).

For a more general description of the available models that can be fitted by these programs, see Chap. 12.

Chapter 3
Extensions of the Rasch Model and Alternative Binary Models

The Rasch model assumes that the item characteristic functions are logistic functions, which is a very specific choice. In particular, in the early days of item response modeling, the normal-ogive model proposed by Lord (1952) has been considered an important alternative. In the following, a wider classes of models is considered. Alternative binary models are also useful when considering models with more than two answer categories since they are contained in most of the current polytomous item response models. An important extension of the Rasch model is the model with varying slopes or discrimination parameters. The model allows that the discriminatory power varies over items and typically shows better fit to the data than the simpler versions. In the following, the functional forms of extended models are specified. It is implicitly assumed that conditional independence of responses given the latent trait holds.

3.1 Homogeneous Monotone Latent Trait Models

A wider class of models that includes the Rasch model as a member is the family of homogeneous monotone models.

> Homogeneous Monotone Latent Trait Model
>
> $$P(Y_{pi} = 1|\theta_p, \delta_i) = F(\theta_p - \delta_i), \tag{3.1}$$
>
> where $F(.)$ is a strictly monotone distribution function.

As in the Rasch model θ_p represents the ability or attitude of person p, and δ_i is the item parameter representing the difficulty in performance tests. The model is homo-

G. Tutz, *A Short Guide to Item Response Theory Models*, Statistics for Social and Behavioral Sciences, https://doi.org/10.1007/978-3-031-87271-6_3

geneous since the item characteristic functions $ICC_i(\theta_p) = P(Y_{pi} = 1|\theta_p, \delta_i) = F(\theta_p - \delta_i)$ all use the same response function $F(.)$.

If one uses the logistic distribution function $F(\eta) = \exp(\eta)/(1 + \exp(\eta))$ one obtains the Rasch model, also called *1PL model* for one parameter logistic model. Another model that has been widely used is the normal-ogive model, in which $F(.)$ is the standard normal distribution function (Lord, 1952). As the logistic distribution function, which defines the Rasch model, the normal distribution function is symmetric and sigmoidal. However, there is no compelling reason why item response functions should be symmetric. Advantages of skewed response functions have been outlined by Samejima (2000), Bazán et al. (2006), and Bolt and Liao (2022).

Examples of homogeneous models are

- the Rasch model with $F(\eta) = \exp(\eta)/(1 + \exp(\eta))$;
- the normal-ogive model with $F(\eta) = \Phi(\eta)$, where $\Phi(.)$ is the standardized normal distribution function;
- the Gumbel model, in which $F(.)$ is the maximum value distribution function $F(\eta) = \exp(-\exp(-\eta))$;
- the Gompertz model, in which $F(.)$ is the minimum value distribution (Gompertz distribution function) $F(\eta) = 1 - \exp(-\exp(x))$;
- Samejima's modified logistic model, $F(\eta) = (\exp(\eta)/(1 + \exp(\eta)))^{\xi}$, where $\xi > 0$ is a fixed exponent (Samejima, 2000).

Figure 3.1 shows the response functions for the maximum (left) and minimum (right) value models with response functions centered at zero. In addition, the logistic response function is shown as dashed line. The maximum value (Gumbel) model shows positive skewness (lower slope above versus below the inflection point of the ICC). Distributions of that form are also called right-skewed or right-tailed. The minimum value (Gompertz) model shows negative skewness (lower slope below versus above the inflection point), it is left-skewed or left-tailed. Samejima's modified logistic model behaves in a similar way having positive skewness if $\xi > 1$ and negative skewness if $\xi < 1$.

A wider class of parametric models called monotone latent variables models has been considered by Holland and Rosenbaum (1986). This general class of models

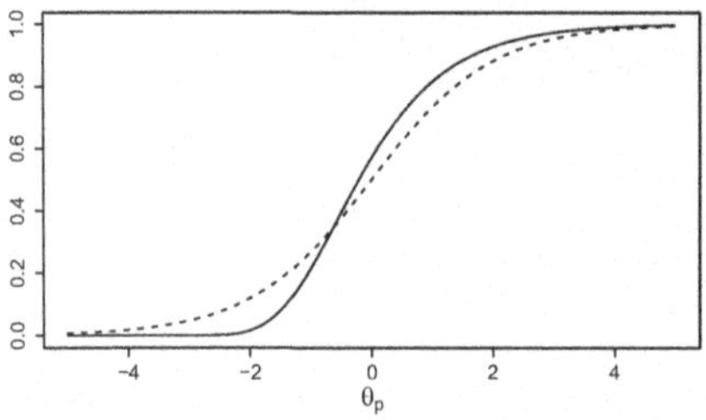

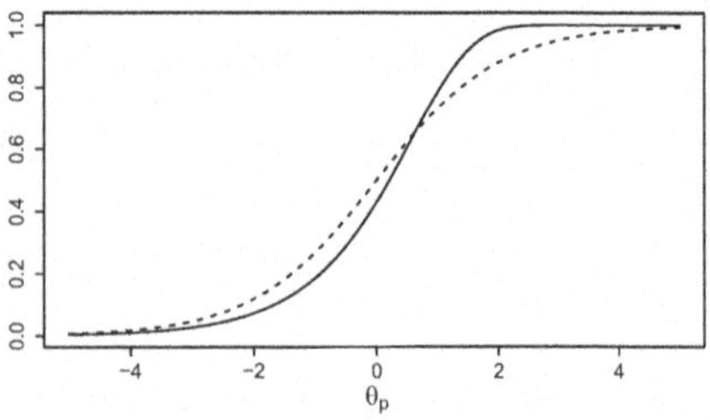

Fig. 3.1 Item characteristic curves centered at zero, left: Gumbel (maximum value distribution), right: Gompertz (minimum value distribution) and logistic model (dashed)

does not restrict items to have identical response functions $F(.)$. Each item can have its own nondecreasing response function. Moreover, the latent trait is not necessarily unidimensional. Holland and Rosenbaum (1986) also outline that conditional independence is not a strong assumption since for responses that have only a finite number of possible values there always exists a unidimensional trait such that conditional independence holds if there is no further restriction of the responses. They clarify that conditions such as monotonicity and functional form are the features that give models testable consequences in observable data.

Another wide class of models based on weak assumptions are so-called nonparametric IRT models also known as Mokken models (Mokken, 1971; Sijtsma and Molenaar, 2002, 2016). They do not assume a fixed functional form for item response functions, response functions are just assumed to be nondecreasing in the latent trait and subjected to order restrictions. Consequently, the likelihood is not fully specified and parameters cannot be estimated in the same way as in parametric models. Instead, estimated expected total scores as in classical test theory are used, making the models more similar to older classical test theory concepts than to latent trait theory modeling approaches.

Model Properties

Homogeneous monotone models have several properties that one wants in binary latent trait models that measure ability or attitude:

They are unidimensional since a one-dimensional person parameter determines the response distribution.

They are monotone since the probability of success increases monotonically with increasing person parameter.

Homogeneous models also yield invariant item ordering: if items are ordered such that $\delta_{[1]} \leq \cdots \leq \delta_{[I]}$ the probabilities of success for the items are ordered by

$$P(Y_{p[1]} = 1|\theta_p, \delta_{[1]}) \geq \cdots \geq P(Y_{p[I]} = 1|\theta_p, \delta_{[I]}) \quad \text{for any } \theta_p. \tag{3.2}$$

Thus, the same ordering of the items by difficulty holds across different levels of the trait or ability.

For more properties of monotone models, see also Holland and Rosenbaum (1986).

Specific Objectivity and Separability of Parameters

In homogeneous models, parameters are separable and a form of specific objectivity holds. It is straightforward to derive that for two persons p_1, p_2

$$F^{-1}(P(Y_{p_1 i} = 1|\theta_{p_1}, \delta_i)) - F^{-1}(P(Y_{p_2 i} = 1|\theta_{p_2}, \delta_i)) = \theta_{p_1} - \theta_{p_2},$$

where $F^{-1}(.)$ is the inverse of $F(.)$. That means that comparison of persons is possible independent of the item involved. In the same way, one has for two items i_1, i_2

$$F^{-1}(P(Y_{pi_1} = 1|\theta_p, \delta_{i_1})) - F^{-1}(P(Y_{pi_2} = 1|\theta_p, \delta_{i_2})) = \delta_{i_2} - \delta_{i_1},$$

which means that items can be compared independent of the persons involved. In the Rasch model, the comparison is especially simple since $F^{-1}(P(Y_{pi} = 1|\theta_p, \delta_i)$ are the log-odds, $F^{-1}(P(Y_{pi} = 1|\theta_p, \delta_i) = \log(P(Y_{pi} = 1|\theta_p, \delta_i)/P(Y_{pi} = 0|\theta_p, \delta_i))$. Although specific objectivity has been considered a specific trait of the Rasch model, it is not the only model for which comparisons are feasible without reference to the measurement instrument if separability refers to separability in the measurement model.

A more general concept of separability in the measurement model uses comparator functions. For a unidimensional latent trait model with response probabilities $\pi_{pi} = P(Y_{pi} = 1)$ for person p and item i *separability of item parameters from person parameters* (*invariance of comparison of item parameters, specific objectivity for the comparison of items*) holds if a parameterization (θ_p, δ_i) and a transformation function C_{it} called comparator exist such that $C_{\text{it}}(\pi_{pi_1}, \pi_{pi_2})$ is equal to the difference of item parameters for all items i_1, i_2. For the parameterization that is assumed to exist, one has

$$C_{\text{it}}(\pi_{pi_1}(\theta_p, \delta_{i_1}), \pi_{pi_2}(\theta_p, \delta_{i_1})) = \delta_{pi_1} - \delta_{pi_2}. \tag{3.3}$$

The definition of separability only assumes that a parameterization exists in order to make the definition independent of the specific parameterization that is used in the formulation of the model. For example, in the Rasch model the parameterization θ_p, δ_i provides such a parameterization but the parameterization θ^*_p, δ^*_i (with $\theta^*_p = \exp(\theta_p), \delta^*_i = \exp(\delta_i)$) does not. It is obvious that homogeneous models are separable in this sense. Of course, a corresponding definition concerning the invariance of comparison of person parameters can be given.

As already emphasized empirical separability is a different concept, it refers to estimation not the probabilistic measurement itself. Empirical separability for homogeneous models is harder to obtain since for non-logistic models no sufficient statistics are available and therefore conditional likelihood estimation cannot be used. For empirical separability by using pairwise partition estimators, see Tutz (2024a).

Goodness of Fit

Binary observations do not contain much information. Thus, models with different response functions typically can only be distinguished if the dataset is rather large. Even then differences in goodness of fit can be rather small as in the following example.

Example 3.1 Fears data

We consider again the fears data from Example 2.4. The response refers to fear concerning five questions ($I = 5$) with fear assessed by a binary response. The number of persons in the full dataset, which is used here, is 2036. Table 3.1 shows the log-likelihoods of the Rasch model and two models with non-symmetric response functions together with the estimated variance of the person parameters (marginal estimation with normal distribution of person parameters). The Rasch model shows the best fit but the fits of the other two models are not much worse. The estimates of the standard deviation of the mixing distribution are quite different, which

Table 3.1 Log-likelihoods for alternative models (fears data)

	Log-likelihood	Standard deviation person parameters
Logistic (Rasch) model	−6215.713	1.196
Gumbel model	−6220.714	0.845
Gompertz model	−6219.378	0.774

is to be expected since the response functions differ in their steepness (the variance of the logistic distribution differs from that of the Gumbel and the Gompertz distribution). □

3.2 Models with Varying Slopes

The class of homogenous monotone models can be made more flexible by using more complex parameterizations that allow the slope and the intercept of the item characteristic functions to vary across items.

Homogeneous Monotone Models with Slopes

$$P(Y_{pi} = 1|\theta_p, \delta_i) = F(\alpha_i(\theta_p - \delta_i)), \alpha_i \geq 0. \tag{3.4}$$

The model has two item parameters, the location parameter δ_i and the discrimination or slope parameter α_i. When using the logistic distribution function the model is often referred to as *2PL model* (Birnbaum, 1968). A special version that uses the normal distribution was considered already by Lord (1952).

The effect of the discrimination parameter is visualized in Fig. 3.2 for the logistic model. It shows two items with differing discrimination parameters, $\alpha_1 = 1$ for

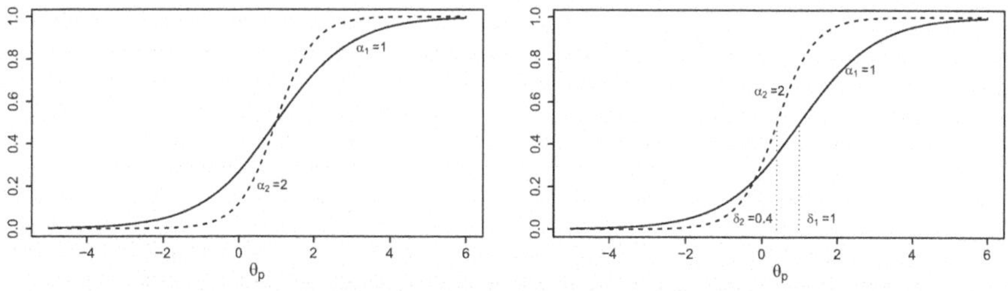

Fig. 3.2 Left: item characteristic curves for two items with the same location parameter ($\delta_1 = \delta_2 = 1$) and differing discrimination parameters $\alpha_1 = 1$ for item 1 (drawn line) and $\alpha_2 = 2$ for item 2 (dashed). Right: item characteristic curves for two items with differing location parameters ($\delta_1 = 1, \delta_2 = 0.4$) and differing discrimination parameters

item 1 (drawn line) and $\alpha_2 = 2$ for item 2 (dashed). The left picture shows the item characteristic curves when the difficulties are the same, in the right picture they differ. It is seen that the item characteristic curves cross if items have differing discrimination parameters. One consequence is larger δ_i does not automatically imply smaller probability of success. In the left picture, item 1 is easier to solve than item 2 if $\theta_p < 1$ but harder to solve if $\theta_p > 1$. Thus, in the model with slopes δ_i does not represent the actual difficulty of the item. It has to be seen in connection with the discrimination parameter, which also determines the probability of success. We refer to δ_i as a *location parameter* since it is on the same scale as the person parameter. It is a difficulty parameter only in the limited sense that it indicates where the probability of success is 0.5 (in the logistic model).

If discrimination parameters vary over items there is no ordering of the items by difficulty across different levels of the trait or ability. Invariant item ordering (3.2), which holds for the model with fixed slopes, $\alpha_i = 1$, and which is often considered desirable, does not hold. In addition, parameters cannot be separated in the simple way that is possible for homogeneous models. For two persons p_1, p_2, one obtains

$$F^{-1}(P(Y_{p_1 i} = 1|\theta_{p_1}, \delta_i)) - F^{-1}(P(Y_{p_2 i} = 1|\theta_{p_2}, \delta_i)) = \alpha_i(\theta_{p_1} - \theta_{p_2}),$$

which contains the item parameter α_i. Thus, the additional flexibility of the model with discrimination parameter comes with a price.

Models with discrimination parameters typically show better fits than models in which the discrimination parameter is fixed. Since the model with discrimination parameters is a submodel of the model without discrimination parameters, likelihood ratio tests can be used to investigate if the discrimination parameters are significantly different from one. The scaled difference in log-likelihoods

$$\lambda = -2 * (l(\text{fitted submodel}) - l(\text{fitted model}))$$

indicates if the submodel can be considered a good approximation of the more general model when compared to the $1 - \alpha$-quantile of the χ-squared distribution with the degrees of freedom given as the difference in the number of parameters in the two models.

An alternative is to use goodness-of-fit measures that account for the number of parameters. A widely used criterion is Akaike's information criterion which he named AIC, for "an information criterion" (Akaike, 1973). It is defined by

$$AIC = -2l(\text{fitted model}) + 2 \cdot (\text{number of fitted parameters}),$$

where $l(\hat{\beta})$ denotes the log-likelihood of the fitted model. The second term may be seen as a penalty accounting for the number of fitted parameters.

Example 3.2 Knowledge Data
In the package *psychotree* (Strobl et al., 2015), the dataset SPISA is included, which contains binary responses from an online quiz for testing one's general knowledge

Table 3.2 Estimates of item parameters for knowledge data, 2PL and 1PL models

	2PL		Rasch	
Item	Difficulty	Discrimination	Difficulty	Discrimination
V1	−2.392	0.825	−2.026	1
V2	−1.331	0.857	−1.165	1
V3	1.155	0.249	0.332	1
V4	−2.579	0.684	−1.877	1
V5	−0.250	1.033	−0.253	1
V6	−2.031	0.731	−1.565	1
V7	−2.489	0.511	−1.410	1
V8	−1.157	0.773	−0.935	1
V9	−2.494	1.096	−2.614	1
log-lik	−5062.843		−5100.601	

conducted by the weekly German news magazine SPIEGEL. The 45 test questions were from five topics, politics, history, economy, culture, and natural sciences. We use the nine items that refer to the natural sciences consisting of 1075 university students from Bavaria. Table 3.2 shows marginal parameter estimates for the 2PL model and the Rasch or 1PL model. It is seen that the 2PL model shows a significantly better fit to the data.

Also Andersen's LR test supports that the Rasch model cannot be expected to be a good representation of the data. When subgroups are defined by the mean (below or above) Andersen LR test yields 31.28 on 8 df. Figure 3.3 shows the corresponding graphical model checks. It is seen that in particular item 3, which is the one with the smallest discrimination parameter, and item 9, which is the one with the largest

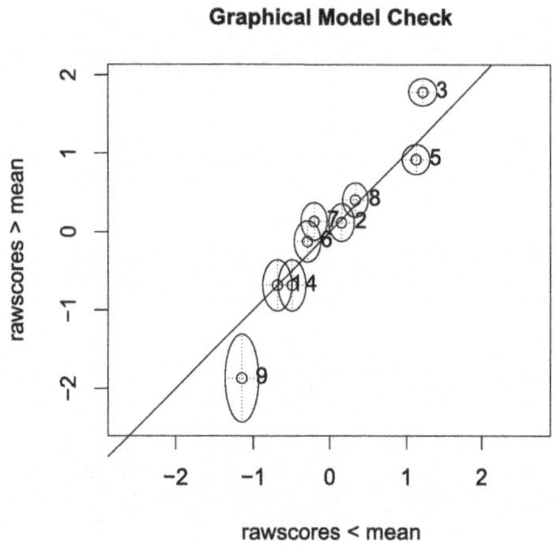

Fig. 3.3 Graphical model check for knowledge data, split criterion is the mean of raw scores

discrimination parameter, seem not to be compatible with the assumption of equal discrimination parameters. □

While models with discrimination parameters show better fits than models in which the discrimination parameter is fixed they are somewhat harder to fit and, as the following example illustrates, sometimes algorithms do not converge properly.

Example 3.3 FIMS study
In the FIMS dataset (Example 2.3), responses on 14 mathematics items are available. Table 3.3 shows marginal parameter estimates for the 2PL model and the Rasch or 1PL model. It is seen that the 2PL model shows much better fit but estimation is hardly reliable since some parameter estimates tend to deteriorate when fitted with function gpcm from package ltm (see Sect. 3.5). In particular, items 12 and 13 have very small discrimination parameters, but also the discrimination parameters of items 2, 7, 8, 9 are distinctly different from 1. If these items are omitted estimates are more reliable and the log-likelihoods of the two models are much more closer (-1172.93 for the Rasch model and -1165.251 for the 2PL model). The corresponding likelihood ratio test yields 15.358 on 8 df (p-value 0.052), which indicates that in the reduced item set the Rasch model could be an approximation that works as well as the more general model. For the AIC criterion, one obtains 2363.861 for the Rasch model and 2364.502 for the 2PL model. Thus, in the reduced item set, the AIC criterion just weakly prefers the Rasch model over the 2PL model (Table 3.4). □

Table 3.3 Estimates of item parameters for FIMS study, 2PL and 1PL models

	2PL		Rasch	
Item	Difficulty	Discrimination	Difficulty	Discrimination
1	−1.107	1.335	−1.318	1
2	−0.755	2.292	−1.157	1
3	−1.361	1.695	−1.856	1
4	−0.255	1.508	−0.319	1
5	2.772	1.031	2.800	1
6	−1.310	1.076	−1.360	1
7	1.300	0.544	0.789	1
8	1.731	0.297	0.600	1
9	3.176	0.502	1.776	1
10	−0.554	1.010	−0.548	1
11	1.524	1.913	2.207	1
12	90.265	0.016	1.676	1
13	15.905	0.135	2.455	1
14	−0.959	1.220	−1.080	1
log-lik	−2070.247		−2126.192	

Table 3.4 Estimates of item parameters for FIMS study, 2PL and 1PL models, reduced item set

	2PL		Rasch	
Item	Difficulty	Discrimination	Difficulty	Discrimination
1	−1.030	1.513	−1.357	1
3	−1.391	1.621	−1.910	1
4	−0.261	1.415	−0.327	1
5	2.887	0.975	2.871	1
6	−1.328	1.053	−1.400	1
10	−0.553	1.011	−0.564	1
11	1.578	1.774	2.269	1
14	−0.917	1.309	−1.112	1
log-lik	−1165.251		−1172.93	

3.3 Three-Parameter Models

The three-parameter model introduced by Birnbaum (1968) can be given in the following form.

Three-Parameter Homogeneous Monotone Models

$$P(Y_{pi} = 1|\theta_p, \delta_i) = c_i + (1 - c_i)F(\alpha_i(\theta_p - \delta_i)), \qquad (3.5)$$

where $\alpha_i \geq 0$, $c_i \in [0, 1)$.

The model has one item parameter more than the model with a discrimination parameter, and is often referred to as 3PL model if for $F(.)$ the logistic distribution function has been chosen. The additional parameter c_i is typically included to account for random guessing which might be present in multiple-choice data. It changes the baseline probability of a response $Y_{pi} = 1$. Figure 3.4 shows the item characteristic curves of the items given in Fig. 2.3 but with a parameter $c_i = 0.2$. It is seen that the probability of solving an item is at least 0.2 for all persons. This corresponds to the case of a multiple-choice item with one correct answer and persons that answer randomly. However, since preferences for specific categories might vary across items the parameter c_i should be estimated instead of being considered a known fixed value.

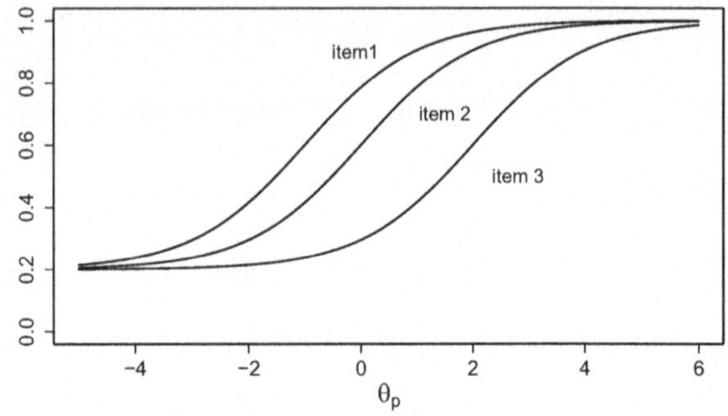

Fig. 3.4 Item characteristic curves for three items with guessing parameter $c_i = 0.2$

3.4 Estimation

For models with general response functions and discrimination parameters in general, there are no sufficient statistics available. An exception is the logistic model, for which sufficient statistics are available if the discrimination parameters are fixed and known, see Verhelst and Glas (1995). However, in practice, discrimination parameters are typically not known. Marginal maximum likelihood estimation has the advantage that sufficient statistics are not needed.

Marginal Maximum Likelihood Estimation
Marginal maximum likelihood (MML) estimates of item parameters are obtained by assuming a fixed distribution of item parameters. The corresponding marginal likelihood has the form:

$$L_m = \prod_{p=1}^{P} \int \prod_{i=1}^{I} P(Y_{pi} = 1)^{y_{pi}} P(Y_{pi} = 0)^{1-y_{pi}} f(\theta_p) d\theta_p,$$

where $f(\theta_p)$ is the density of the distribution assumed for the person parameters. Typically, one assumes a normal distribution with mean zero and variance σ_θ^2, which can be abbreviated by $f_{0,\sigma_\theta}(.)$. The corresponding log-likelihood is

$$\begin{aligned} l_m = \log(L_m) &= \sum_{p=1}^{P} \log(\int \prod_{i=1}^{I} P(Y_{pi} = 1)^{y_{pi}} ((1 - c_i) P(Y_{pi} = 0)^{1-y_{pi}} f_{0,\sigma_\theta} d\theta_p), \\ &= \sum_{p=1}^{P} \log(\int \prod_{i=1}^{I} f_{pi}(y_{pi}) f_{0,\sigma_\theta}(\theta_p) d\theta_p), \end{aligned}$$

where $f_{pi}(y_{pi}) = (c_i + (1 - c_i) F(\alpha_i(\theta_p - \delta_i))^{y_{pi}} ((1 - c_i)(1 - F(\alpha_i(\theta_p - \delta_i)))^{1-y_{pi}}$.

The score function $s(\boldsymbol{\delta}) = \partial l/\partial \boldsymbol{\delta}$, which takes on the value of 0 for maximum likelihood estimates, has components

$$\frac{\partial l}{\partial \delta_i} = \sum_{p=1}^{P} \int \frac{\partial f_{pi}(y_{pi})}{\partial \delta_i} \prod_{l \neq i} f_{pi}(y_{pl}) f_{0,\sigma_\theta}(\theta_p) d\theta_p / \gamma_p, \quad \frac{\partial f_{pi}(y_{pi})}{\partial \delta_i} = (-1)^{y_{pi}} \alpha_i f(\alpha_i(\theta_p - \delta_i)),$$

$$\frac{\partial l}{\partial \alpha_i} = \sum_{p=1}^{P} \int \frac{\partial f_{pi}(y_{pi})}{\partial \alpha_i} \prod_{l \neq i} f_{pi}(y_{pl}) f_{0,\sigma_\theta}(\theta_p) d\theta_p / \gamma_p, \quad \frac{\partial f_{pi}(y_{pi})}{\partial \alpha_i} = (-1)^{(1-y_{pi})}(\theta_p - \delta_i) f(\alpha_i(\theta_p - \delta_i)),$$

$$\frac{\partial l}{\partial \sigma_\theta} = \sum_{p=1}^{P} \int \prod_{i=1}^{I} f_{pi}(y_{pi}) \frac{\partial f_{0,\sigma_\theta}(\theta_p)}{\partial \sigma_\theta} d\theta_p / \gamma_p,$$

where $\gamma_p = \int \prod_{i=1}^{I} f_{pi}(y_{pi}) f_{0,\sigma_\theta}(\theta_p) d\theta_p$ and $f(\eta) = \partial F(\eta)/\partial \eta$ denotes the density that is linked to $F(.)$.

Estimates of the general model with discrimination and guessing parameters can become unstable, a problem that can be solved by using regularized estimation methods, see, for example, Battauz and Bellio (2021). Pairwise estimates for the 2PL model have been propagated more recently by Robitzsch (2024).

Bayesian Approaches

With the advent of powerful computational simulation methods Bayesian estimation approaches have become more popular. They can be very useful in more complex fitting problems and typically avoid problems with the existence of estimates. In addition, Bayesian approaches allow to make inferences without relying on asymptotic distributions. A potential downside is that prior distributions have to be chosen. In the following, basic Bayesian estimation concepts are considered only briefly.

Let very generally $\boldsymbol{y}$ denote the realization of a random vector and $\boldsymbol{\xi}$ a vector of parameters. Let $p(\boldsymbol{y}|\boldsymbol{\xi})$ denote the probability of observations given the parameters, which is equivalent to the likelihood when considered a function of the parameters, and $p(\boldsymbol{\xi})$ the prior distribution (density) of parameters that is assumed. Then the posterior density $p(\boldsymbol{\xi}|\boldsymbol{y})$ of the parameter given prior beliefs $p(\boldsymbol{\xi})$ and sample information contained in $p(\boldsymbol{y}|\boldsymbol{\xi})$ is given by

$$p(\boldsymbol{\xi}|\boldsymbol{y}) = \frac{p(\boldsymbol{y}|\boldsymbol{\xi})p(\boldsymbol{\xi})}{p(\boldsymbol{y})} \propto p(\boldsymbol{y}|\boldsymbol{\xi})p(\boldsymbol{\xi}),$$

where $p(\boldsymbol{y})$ is the probability of observations. In Bayesian approaches, the posterior $p(\boldsymbol{\xi}|\boldsymbol{y})$ is used to make inferences.

Bayesian approaches can be used to obtain estimates for person and item parameters simultaneously and for person parameters only or item parameters only within the framework of Bayesian hierarchical modeling (Fox, 2010).

When estimating, for example, item parameters one uses the marginal posterior distribution $p(\gamma|\boldsymbol{y})$. The posterior assuming independence of item and person parameters is given by

$$p(\boldsymbol{\theta}, \gamma|\boldsymbol{y}) = p(\boldsymbol{y}|\boldsymbol{\theta}, \gamma)p(\boldsymbol{\theta}, \gamma)/p(\boldsymbol{y}) = p(\boldsymbol{y}|\boldsymbol{\theta}, \gamma)p(\boldsymbol{\theta})p(\gamma)/p(\boldsymbol{y}).$$

Let $\boldsymbol{y}_{.i}^T = (y_{1i}, \ldots, y_{Pi})$ denote the vector of responses on item i, $\boldsymbol{\gamma}^T = (\gamma_1, \ldots, \gamma_I)$ the item parameters and $\boldsymbol{\delta}^T = (\delta_1, \ldots, \delta_P)$ the person parameters. Then, the marginal posterior of item parameters is given by

$$p(\boldsymbol{\gamma}|\boldsymbol{y}) = \int p(\boldsymbol{\theta}, \boldsymbol{\gamma}|\boldsymbol{y})d\boldsymbol{\theta} = \frac{1}{p(\boldsymbol{y})}\int p(\boldsymbol{y}|\boldsymbol{\theta}, \boldsymbol{\gamma})p(\boldsymbol{\theta})p(\boldsymbol{\gamma})d\boldsymbol{\theta} =$$

$$= \frac{1}{p(\boldsymbol{y})}\int \prod_{i=1}^{I} p(\boldsymbol{y}_{.i}|\boldsymbol{\theta}, \gamma_i)p(\boldsymbol{\theta})p(\boldsymbol{\gamma})d\boldsymbol{\theta}.$$

Assuming that the item parameters are independent conditional on the hyperparameter vector γ_H and identically distributed one obtains with $p(\boldsymbol{\gamma}) = p(\gamma_1|\gamma_H)\ldots p(\gamma_I|\gamma_H)p(\gamma_H)$

$$p(\boldsymbol{\gamma}|\boldsymbol{y}) = \frac{1}{p(\boldsymbol{y})}\prod_{i=1}^{I}\int p(\boldsymbol{y}_{.i}|\boldsymbol{\theta}, \gamma_i)p(\gamma_i)p(\gamma_i|\gamma_H)p(\gamma_H)p(\boldsymbol{\theta})d\boldsymbol{\theta}.$$

Bayesian estimates of item parameters can be derived from $p(\boldsymbol{\gamma}|\boldsymbol{y})$ after specifying prior distributions, which should be done carefully since estimates depend on that choice. An extensive overview on Bayesian item response modeling has been given by Fox (2010), see also Van der Linden (2016a).

3.5 Fitting of Models

Marginal estimates of the 2PL model for the FIMS data can be obtained by using the R package *ltm* (Rizopoulos, 2006). The parameter estimates for the 2PL model given in Table 3.2 can be obtained by using the following code.

```
library(psychotree)
data(SPISA)
SP <- as.data.frame(SPISA[,1])
SPNat <- SP[,37:45]
library("ltm")
pcmltm<-gpcm(SPNat, constraint = "rasch")
pcmltm
pcmltmd<-gpcm(SPNat, constraint = "gpcm")
pcmltmd
```

The parameter estimates for the 2PL model given in Table 3.3 can be obtained by using the following code.

```
library("ltm")
datscored<-readRDS("datAust300.RDS")
pcmltmd<-gpcm(datscored, constraint = "gpcm")
pcmltmd
pcmltmdR<-gpcm(datscored, constraint = "rasch")
pcmltmdR
```

pcmltmd gives the estimates for the 2PL model and pcmltmdR for the Rasch model.

Marginal estimates of the 2PL model can also be obtained by using the R package TAM (Robitzsch et al., 2024). The parameter estimates for the 2PL model given in Table 3.2 can be obtained by using the following code.

```
library(TAM)
mod1 <- TAM::tam.mml(resp=datscoredred,irtmodel="1PL")
summary(mod1)
mod2 <- TAM::tam.mml.2pl(resp=datscoredred, irtmodel="2PL")
summary(mod2)
```

The parameter estimates for the reduced item set can be obtained by using the following code. Fitting of the reduced item set 1, 3, 4, 5, 6, 10, 11, 14 is obtained by the following.

```
sel<-c(1,3,4,5,6,10,11,14)
datscoredred<- datscored [,sel]
pcmltmred<-gpcm(datscoredred, constraint = "rasch")
pcmltmred21<-gpcm(datscoredred, constraint = "gpcm")
pcmltmred
pcmltmred21
```

TAM yields the same estimates for the 2PL model and slightly different estimates for the Rasch model.

Chapter 4
Ordinal Models

In many applications, items have more than two response categories with categories being ordered. Let polytomously scored items have values $Y_{pi} \in \{0, 1, \dots, k\}$, $p = 1, \dots, P$, $i = 1, \dots, I$, with $\{0, 1, \dots, k\}$ representing the ordering of categories. For simplicity we use a fixed number of response categories when defining the models. In general, the number of response categories can vary across items with the response categories of item i given by $\{0, 1, \dots, k_i\}$.

In achievement tests, the response categories $0, 1, \dots, k$ refer to levels of performance, and in attitude tests they indicate the degree of accordance with a statement. A specific type of item that in the social and behavioral sciences has been frequently used to measure attitudes, character, and personality traits are Likert scales, which raise some interesting modeling problems because of their symmetric nature. Following their introduction by Likert (1932) various versions have been used, the most popular versions are five-grade and seven-grade Likert scales. The grades $0, \dots, k$ are assumed to be ordered and reflect agreement/disagreement or approval/disapproval of the respondent with respect to the value statement. In five-grade Likert scales, the grades are typically interpreted by strongly disagree, disagree, neutral (undecided), agree, and strongly agree. One may distinguish between Likert-type items and Likert scales, see, for example, Kaptein et al. (2010). The term Likert-type item is typically used if individual items are considered as measurement tools, the term Likert scale if scores are computed from a number of relating items, however in the following we will not use this distinction.

4.1 Binary Models as Building Blocks for Ordinal Models

All ordinal models have to distinguish between low categories which in achievement tests represent weak performance and high categories which indicate high performance. The same holds for attitude tests where low numbers refer to weak agreement

G. Tutz, *A Short Guide to Item Response Theory Models*, Statistics for Social and Behavioral Sciences, https://doi.org/10.1007/978-3-031-87271-6_4

Table 4.1 Types of ordinal models for response $Y \in \{0, 1, \dots, k\}$ and dichotomous variables $Y^{(r)}$ that are contained

Cumulative-Type Model, Dichotomization into Groups	
$[0, \dots, r-1 \vert r, \dots, k]$	$Y^{(r)} = \begin{cases} 1 & Y \geq r \\ 0 & Y < r \end{cases}$
Adjacent-Type Model, Dichotomization given $Y \in \{r, r+1\}$	
$0, \dots, [r-1 \vert r], \dots, k$	$Y^{(r)} = \begin{cases} 1 & Y \geq r \\ 0 & Y < r \end{cases}$ given $Y \in \{r-1, r\}$
Sequential-Type Model, Dichotomization given $Y \geq r$	
$0, \dots, [r-1 \vert r, \dots, k]$	$Y^{(r)} = \begin{cases} 1 & Y \geq r \\ 0 & Y < r \end{cases}$ given $Y \geq r-1$

and high numbers refer to strong agreement. Consequently ordinal models can be constructed from binary models that distinguish between weak and strong performance. The three major types of ordinal models that are in common use can be simply characterized by the way they use binary models as building blocks.

Given one has a response in *ordered* categories $\{0, 1, \dots, k\}$ there are several ways to construct an ordinal model from binary models. The binary models can be used to compare specific categories or groups of categories from $\{0, 1, \dots, k\}$. One can, in particular,

- compare groups of categories that result from splitting the categories into the subsets $\{0, 1, \dots, r-1\}$ and $\{r, \dots, k\}$;
- compare (conditionally) between two categories, for example, adjacent categories $\{r-1, r\}$;
- compare (conditionally) between a category and a set of adjacent categories, for example, $\{r-1\}$ and $\{r, \dots, k\}$.

The different ways to compare categories correspond to the most widely used ordinal models, which are cumulative, adjacent categories, and sequential models.

Table 4.1 shows three types of dichotomizations that determine these models. With an ordinal response Y taking values $0, 1, \dots, k$, the variables $Y^{(r)}$ denote the binary variables that are contained in the model. Brackets denote which response categories are used when dichotomizing and the "|" determines the (conditional) split. The table shows which binary models are contained in ordered latent trait models but can also be used to construct models from binary variables. Figure 4.1 distinguishes between conditional and non-conditional models. The cumulative or graded response model is the only non-conditional model from this class of models. The other two use some sort of conditioning when constructing ordinal models from binary ones.

Representation of Polytomous Items by Split Variables

The splitting of categories into groups of adjacent categories, which distinguishes between weak and strong performance, can in general be described by *split variables*, which are defined by

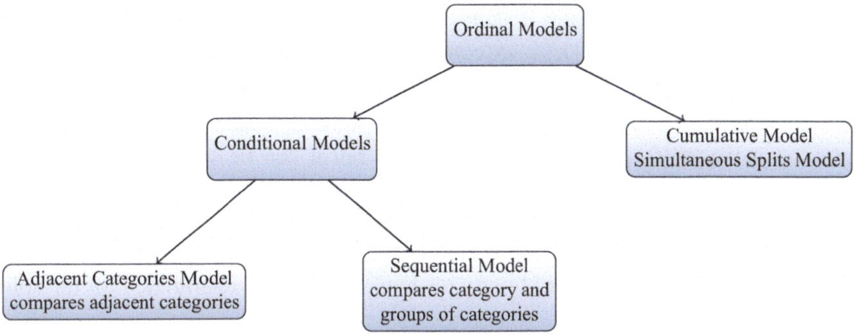

Fig. 4.1 Structure of classical ordinal latent trait models

Table 4.2 Response $Y_{pi} \in \{0, 1, 2, 3\}$ and representation by split variables

Y_{pi}	$(Y_{ci}^{(1)},$	$Y_{pi}^{(2)},$	$Y_{pi}^{(3)})$
0	(0,	0,	0)
1	(1,	0,	0)
2	(1,	1,	0)
3	(1,	1,	1)

$$Y_{pi}^{(r)} = \begin{cases} 1 & Y_{pi} \geq r \\ 0 & Y_{pi} < r. \end{cases} \tag{4.1}$$

The split variables are binary variables that partition the categories into the subsets $\{0, \ldots, r-1\}$ and $\{r, \ldots, k\}$ using a specific definition of weak and strong performance. With response categories $\{0, 1, \ldots, k\}$, one has k split variables $Y_{pi}^{(1)}, \ldots, Y_{pi}^{(k)}$. Split variables are also interesting because there is a one-to-one correspondence of split variables and the categorical response given by

$$Y_{pi} = r \quad \Leftrightarrow \quad (Y_{pi}^{(1)}, \ldots, Y_{pi}^{(r)}, Y_{pi}^{(r+1)}, \ldots, Y_{pi}^{(k)}) = (1, \ldots, 1, 0, \ldots, 0), \tag{4.2}$$

where $Y_{pi}^{(r)}$ is the last of the sequence of binary variables with a value 1. The sequence $Y_{pi}^{(1)} \ldots, Y_{pi}^{(k)}$ can be seen as a specific coding of the multi-categorical response Y_{pi}. For illustration, Table 4.2 shows the response values and the corresponding coding by split variables for four response categories.

Thus, each observed polytomous response can be coded as a vector of split variables. To each response there is a unique vector of split variables and vice versa. The polytomous response can be simply obtained from the split variable representation by

$$Y_{pi} = \sum_{j=1}^{k} Y_{pi}^{(j)}.$$

The sequences $(Y_{pi}^{(1)} \ldots, Y_{pi}^{(k)})$ of split variables have the special property that a sequence of ones is followed by a sequence of zeros. Random variables $Y_{pi}^{(1)} \ldots, Y_{pi}^{(k)}$ with $Y_{pi}^{(r)} \in \{0, 1\}$ and $Y_{pi}^{(r)} \geq Y_{pi}^{(r+1)}$ have been called *Guttman variables* (they have a *Guttman structure*). In general, the Guttman space is defined by $G = \{(y_1, \ldots, y_k)|y_r \in \{0, 1\}, y_r \geq y_s$ for all $r < s\}$ but it can be simply thought of as the vector-valued representation of responses in categories generated by split variables. It should be noted that the representation of polytomously scored items as vectors of split variables holds in general, no ordering of categories is assumed. The order comes into play when binary response models are formulated for the split variables.

In the following classical ordinal models will be represented as models for split variables, which makes the models and the interpretation of parameters easily accessible.

4.2 Graded Response Models

One of the most widely used ordinal model is Samejima's *graded response model* (Samejima, 1969, 1995, 2016). It can be seen as being composed from unconditional binary models for the splits between categories.

Let the response categories represent levels of performance in an achievement test and consider two groups of categories, $\{0, 1, \ldots, r-1\}$ for low performance and $\{r, \ldots, k\}$ for high performance, where low and high are relative terms that refer to "below category r" and "above or in category r". Let the split into low and high performance be determined by a binary model with person ability θ_p and a threshold δ_{ir} that depends on the category at which the categories have been split by specifying

$$P(Y_{pi}^{(r)} = 1|\theta_p, \alpha_i, \delta_{ir}) = F(\alpha_i(\theta_p - \delta_{ir},)) \quad , r = 1, \ldots, k. \tag{4.3}$$

Thus, for each split variable, that is, the dichotomization into categories $\{0, 1, \ldots, r-1\}$ and $\{r, \ldots, k\}$, a binary model is assumed to hold. Importantly, the models are assumed to hold *simultaneously* with the same person ability θ_p but different location parameters δ_{ir}. Simple rewriting yields the graded response model, which is also called *cumulative model.*

Graded Response Model or Cumulative Model

$$P(Y_{pi} \geq r|\theta_p, \alpha_i, \boldsymbol{\delta}_i) = F(\alpha_i(\theta_p - \delta_{ir})), \quad r = 1, \ldots, k. \tag{4.4}$$

In the definition, the location parameters are collected in the vector $\boldsymbol{\delta}_i^T = (\delta_{i1}, \ldots, \delta_{ik})$. The model has originally been introduced by Samejima as a normal-ogive model, in which $F(.)$ is the normal distribution. The logistic version uses the logistic function $F(.)$.

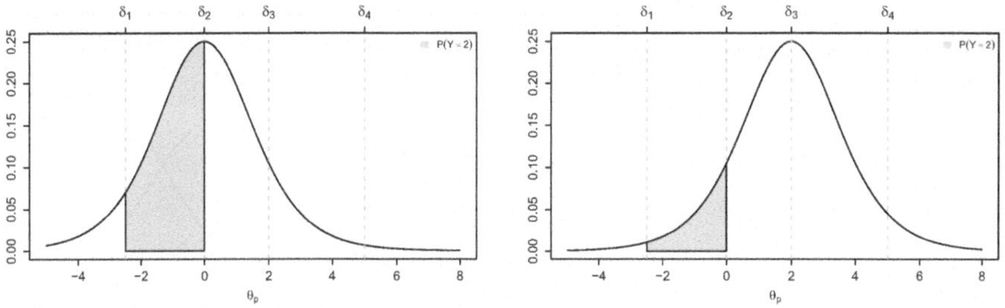

Fig. 4.2 Densities of the random ability for persons $\theta_p = 0$ (left) and $\theta_p = 2$, thresholds are $(\delta_{i1}, \ldots, \delta_{ik}) = (-2.5, 0, 2, 5)$, $k = 4$, and $\alpha_i = 1$

The model as presented here is constructed from binary models for the dichotomizations into the categories $Y_{pi} < r$ and $Y_{pi} \geq r$. One consequence is that location parameters are ordered. Since

$$P(Y_{pi} = r) = P(Y_{pi} \geq r) - P(Y_{pi} \geq r + 1) = F(\alpha_i(\theta_p - \delta_{ir})) - F(\alpha_i(\theta_p - \delta_{i,r+1})) \geq 0,$$

one obtains that $\delta_{ir} \leq \delta_{i,r+1}$ has to hold for all categories.

The construction from binary models is simple and straightforward. However, the model can also be derived in a quite different way that suggests that the item parameters may be seen as thresholds. Let $Y^*_{pi} = \theta_p + \sigma_i \varepsilon_{pi}$, where ε_{pi} is a noise variable with symmetric continuous distribution function $F(.)$ and $\sigma_i > 0$ denote a latent variable that is invoked if person p tries to solve item i. Y^*_{pi} is essentially the random ability of the person, that is, the ability plus a noise variable. The *category boundaries approach*, which dates back at least to Edwards and Thurstone (1952), assumes that category r is observed if the latent variable is between thresholds δ_{ir} and $\delta_{i,r+1}$. More formally, one has

$$Y_{pi} = r \quad \Leftrightarrow \quad \delta_{ir} \leq Y^*_{pi} < \delta_{i,r+1}.$$

It is easily seen that one obtains the graded response or cumulative model with $\alpha_i = 1/\sigma_i$. Of course, thresholds have to be ordered, $\delta_{i1} \leq \cdots \leq \delta_{ik}$, otherwise the category boundaries approach would not make sense.

Figure 4.2 shows the densities of the random ability for persons $\theta_p = 0$ (left) and $\theta_p = 2$ (right), thresholds are $(\delta_{i1}, \ldots, \delta_{ik}) = (-2.5, 0, 2, 5)$, $k = 4$, and $\alpha_i = 1$. The probability $P(Y_{pi} = 1)$ is given in gray. It is seen that it is much lower for $\theta_p = 2$ than for $\theta_p = 0$ since the transition from $\theta_p = 0$ to $\theta_p = 2$ means that the density is shifted to the right, and the probability for low categories like 1 decreases. Figure 4.3 shows the corresponding response probabilities as a function of θ_p. It is seen that the probability of a response in category 3 can take higher values than the probabilities of a response in categories 1 and 2, which is due to the chosen thresholds. The interval length $\delta_{i4} - \delta_{i3} = 3$ (category 3) is wider than the intervals $\delta_{i2} - \delta_{i1} = 2.5$ (category 1) and $\delta_{i3} - \delta_{i2} = 2$ (category 2). The difference in adjacent thresholds reflects the

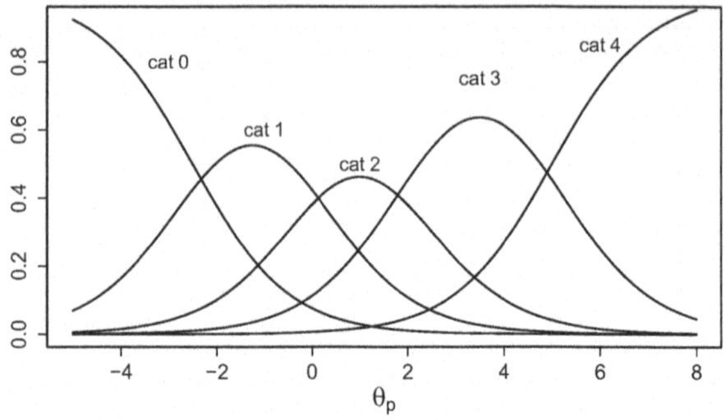

Fig. 4.3 Response probabilities against $\theta_p = 0$ for $k = 4$, $(\delta_{i1}, \ldots, \delta_{ir}) = (-2.5, 0, 2, 5)$, $\alpha_i = 1$

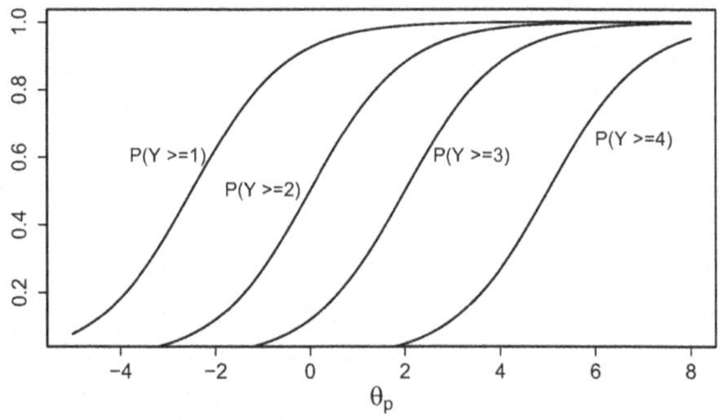

Fig. 4.4 Item characteristic curves for $k = 4$, $(\delta_{i1}, \ldots, \delta_{ik}) = (-2.5, 0, 2, 5)$, $\alpha_i = 1$

preference for specific categories, larger differences indicate a tendency to higher probabilities of the corresponding category.

4.2.1 Properties of Graded Response Models

Item Characteristic Curves

In binary models, the item characteristic curve specifies the probability of success as the level of the ability increases. When responses to items are in more than two categories one obtains more than one item characteristic curve. Let the item characteristic function for item i and response level r be defined by

$$ICC_{i,r}(\theta_p) = P(Y_{pi} \geq r|\theta_p, \boldsymbol{\delta}_i, \alpha_i) = F(\alpha_i(\theta_p - \delta_{ir})). \tag{4.5}$$

Thus for the $k + 1$ response categories $0, 1, \ldots, k$ one obtains k meaningful functions $ICC_{i,1}(.), \ldots, ICC_{i,k}(.)$. Figure 4.4 shows the item characteristic curves for the items considered in Fig. 4.3. Each curve represents the ICC of a binary model, more concise $ICC_{i,r}(.)$ is the ICC of the binary model for the split variable $Y_{pi}^{(r)}$, which splits categories into the subsets $\{0, \ldots, r - 1\}$ and $\{r, \ldots, k\}$.

Information Function

In general, item information (or Fisher information) for item i is defined as the expectation

$$I(\theta_p) = \mathrm{E}\left(-\frac{\partial^2 l_i(Y;\theta_p)}{\partial\theta_p\partial\theta_p}\right),$$

where $l_i(Y;\theta_p)$ is the log-likelihood for item i and observation Y. For discrete $Y \in \{0, 1, \dots\}$ the log-likelihood is given by $l_i(Y;\theta_p) = \sum_r Y_r \log(\pi_{ir}(\theta_p))$, where $Y_r = 1$ if $Y = r$, $Y_r = 0$ otherwise, and $\pi_{ir}(\theta_p) = P(Y_{pi} = r|\theta_p, \alpha_i, \boldsymbol{\delta}_i)$ is the probability of a response in category r on item i. One obtains

$$I(\theta_p) = \sum_r \frac{\pi'_{ir}(\theta_p)^2 - \pi_{ir}(\theta_p)\pi''_{ir}(\theta_p)}{\pi_{ir}(\theta_p)}, \tag{4.6}$$

where $\pi'_{ir}(\theta_p)$ and $\pi''_{ir}(\theta_p)$ are the first and second derivatives of $\pi_{ir}(\theta_p)$ with respect to θ_p (Tutz, 2022b). The derivatives are given by

$$\pi'_{ir}(\theta_p) = \alpha_i F'(\alpha_i(\theta_p - \delta_{ir})) - \alpha_i F'(\alpha_i(\theta_p - \delta_{i,r+1})),$$
$$\pi''_{ir}(\theta_p) = \alpha_i^2 F''(\alpha_i(\theta_p - \delta_{ir})) - \alpha_i^2 F''(\alpha_i(\theta_p - \delta_{i,r+1})),$$

with the derivatives in the logistic model given by $F'(\eta) = F(\eta)(1 - F(\eta))$, $F''(\eta) = F(\eta)(1 - F(\eta))(1 - 2F(\eta))$.

If the number of categories is finite, $Y \in \{0, \dots, k\}$, a simpler form can be derived, in which second derivatives are not needed. By using $\pi_{i0}(\theta_p) = 1 - \pi_{i1}(\theta_p) - \dots - \pi_{ik}(\theta_p)$, one obtains

$$I(\theta_p) = \sum_{r=1}^{k} \frac{\pi'_{ir}(\theta_p)^2}{\pi_{ir}(\theta_p)} + \frac{(\pi'_{i1}(\theta_p) + \dots + \pi'_{ik}(\theta_p))^2}{1 - \pi_{i1}(\theta_p) - \dots - \pi_{ik}(\theta_p)}. \tag{4.7}$$

Figure 4.5 shows the information function for several values of α_i together with the corresponding probability functions for an item with parameters $(\delta_{i1}, \dots, \delta_{i4}) = (-2.5, 0.0, 2.0, 5.0)$. It is seen that information functions are not necessarily concave. While it is concave if the discrimination parameter α_i is small, for larger values the information function becomes wiggly. If, for example, $\alpha_i = 3$ the information can obtain very high values but only if the outcome is uncertain. If the probability for one category is close to 1 the information tends to be very small.

Separability of Parameters

Binary models with fixed discrimination parameters allow to compare items (persons) independently of persons (items). This separability of parameters also called *specific objectivity* holds also for specific graded response models. Let $F(.)$ be the logistic function and $\alpha_i = 1$ for all items. Then one obtains for the *cumulative odds*

$$\gamma_{pi} = \frac{P(Y_{pi} \geq r|\theta_p, \boldsymbol{\delta}_i)}{P(Y_{pi} < r|\theta_p, \boldsymbol{\delta}_i)} = \exp(\theta_p - \delta_{ir}),$$

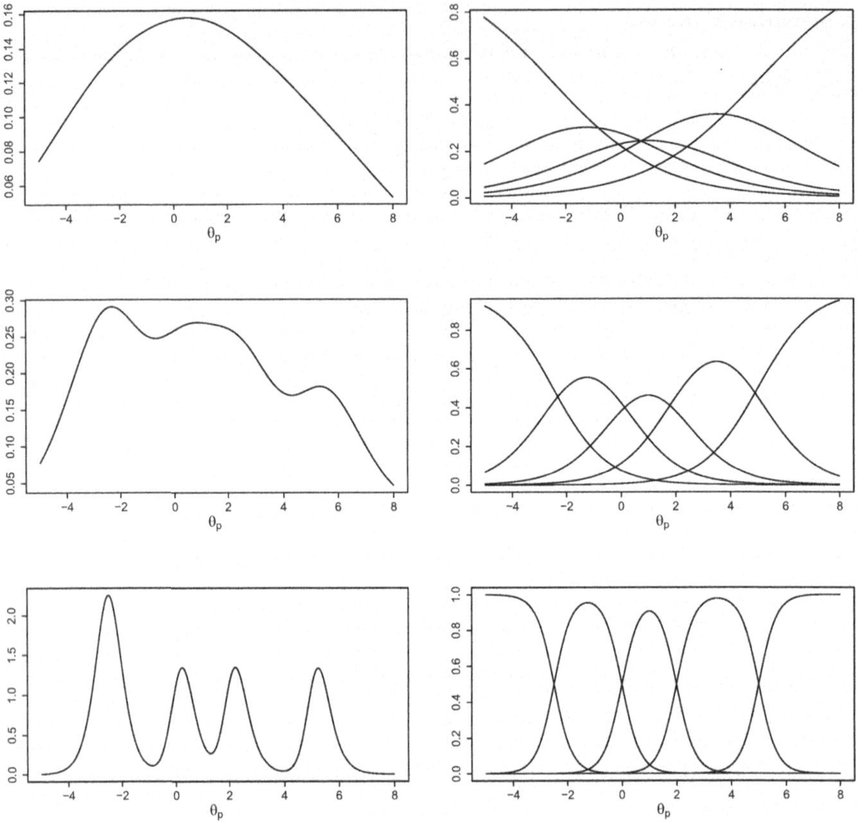

Fig. 4.5 Information functions (left) and response probabilities (right) for $\alpha_i = 0.5$ (first row), $\alpha_i = 1$ (second row), $\alpha_i = 3$ (third row), thresholds are $(\delta_{i1}, \ldots, \delta_{i4}) = (-2.5, 0.0, 2.0, 5.0)$, $k = 4$, and $\alpha_i = 1$

and the cumulative odds ratio for two persons p_1, p_2 is given by

$$\frac{\gamma_{p_1 i}}{\gamma_{p_2 i}} = \frac{\exp(\theta_{p_1})}{\exp(\theta_{p_2})},$$

which does not depend on the item. Thus, comparison of persons is possible independent of the item involved. In the same way, the cumulative odds ratio for two items does not depend on the person. The property does hold not only for the logistic model but for all models with a strictly increasing function $F(.)$. Then the cumulative odds have to be replaced by $F^{-1}(P(Y_{pi} \geq r|\theta_p, \boldsymbol{\delta}_i))$. The crucial assumption to obtain specific objectivity is that discrimination parameters are fixed, that is, $\alpha_i = 1$.

Table 4.3 Observed counts for items

Item	Categories					
	0	1	2	3	4	5
2	3	11	10	145	5	26
3	1	165	9	5	6	14
4	1	3	37	1	109	49
6	5	11	13	149	13	9
7	14	65	14	74	7	26
8	8	68	101	12	8	3

Example 4.1 FIMS study
The FIMS dataset (Example 2.3) contains responses on 14 mathematics items. Here we use a sample of 200 persons from the version with responses from 0 to 5 (data.fims.Aus.Jpn.raw). In the following, we consider a selection of 6 items (items 2, 3, 4, 6, 7, 8). As seen from Table 4.3 for these items each category has been observed at least once. Marginal estimates of the graded response model can be obtained by using the R package *ltm* (Rizopoulos, 2006). Estimates for the reduced item set can be obtained by the following:

```
library("ltm")
datraw<-readRDS("datAust200poly.RDS")
### reduced item set
itemsel<-c(2,3,4,6,7,8)
fitgrad<- grm(datraw[,itemsel], constrained = TRUE, IRT.param = TRUE,
 Hessian = FALSE,start.val = NULL, na.action = NULL, control = list())
fitgrad$coefficients
deltgrad<-as.data.frame(fitgrad$coefficients)
round(t(deltgrad),digits=3)
summary(fitgrad)
```

The package *ltm* fits the models with discrimination parameter α_i. The specification *constrained* = *TRUE* indicates that a model with equal discrimination parameters ($\alpha_i = \alpha$) is fitted. The parameterization that is used is $\eta_{pi} = \alpha_i(\theta_p - \delta_{ir})$, which can be rewritten as $\eta_{pi} = \alpha_i\theta_p - \alpha_i\delta_{ir} = \alpha_i\theta_p - \beta_{ir}$. The coefficients in the list *coefficients* are the parameters $\beta_{ir} = \alpha_i\delta_{ir}$ and are given in Table 4.4.

Figure 4.6 shows the probabilities for items 2 and 6 with scaling $\alpha = 1$. It is seen that some categories have a very low probability for all person parameters, which reflects the low counts seen in Table 4.3. When fitting the generalized model with discrimination parameters (using package ltm) some of the estimated discrimination parameters were negative, which indicates that these items need special attention. □

Table 4.4 Estimated item parameters for logistic graded response model (six items)

Item	$\alpha\delta_{i1}$	$\alpha\delta_{i2}$	$\alpha\delta_{i3}$	$\alpha\delta_{i4}$	$\alpha\delta_{i5}$	α
2	−4.209	−2.610	−2.013	1.710	1.917	0.218
3	−5.315	1.601	1.963	2.216	2.607	0.218
4	−5.315	−3.913	−1.369	−1.338	1.137	0.218
6	−3.686	−2.462	−1.791	2.109	3.077	0.218
7	−2.604	−0.429	−0.140	1.638	1.919	0.218
8	−3.199	−0.493	2.061	2.866	4.208	0.218

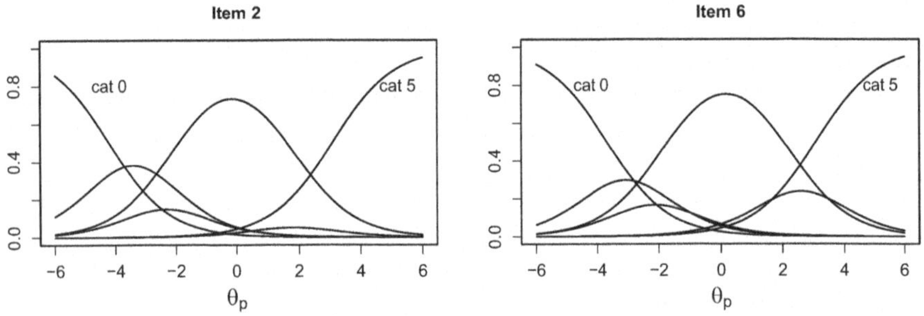

Fig. 4.6 Estimated response probabilities for two items (graded response model)

4.2.2 *Sparser Location and Scale Models**

Models with fewer parameters are always to be preferred since they typically allow simpler interpretation of parameters. In particular, if response formats are symmetrically structured simpler models might be more appropriate. Symmetrically structured response formats are often found in attitude measurement when Likert-type scales are used, which represent the level of agreement in the form *strongly disagree, moderately disagree,..., moderately agree, strongly agree.*

For this type of responses one might assume that thresholds are symmetric. Let the item parameters be reparameterized by

$$\delta_{ir} = \delta_i + \tau_{ir}, \quad r = 1, \ldots, k,$$

where $\sum_l \tau_{il} = 0$. The parameter δ_i is the basic location parameter of the item while the τ-parameters are deviations from it. Then, thresholds are symmetric, centered at δ_i if one assumes

$$\tau_{ij} = -\tau_{i,k+1-j}, \quad j = 1, \ldots, k. \tag{4.8}$$

If $k = 4$ one obtains $\tau_{i1} = -\tau_{i4}$, $\tau_{i2} = -\tau_{i3}$. If $k = 3$ one obtains $\tau_{i1} = -\tau_{i3}$, $\tau_{i2} = 0$. If k is odd the middle threshold is always zero. The assumption of symmetric

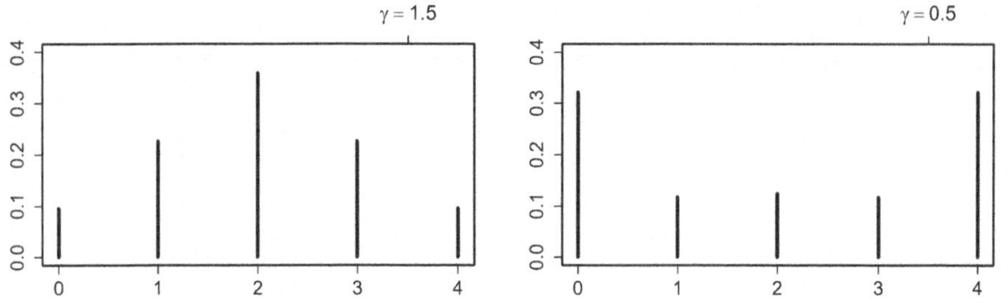

Fig. 4.7 Response probabilities for cumulative logit model with $k = 4$ and two values of γ_i, ($\alpha_i = 1$)

thresholds (4.8) reduces the number of item parameters to $\delta_i, \tau_{i1}, \ldots, \tau_{i,k/2}$ if k is even, and $\delta_i, \tau_{i1}, \ldots, \tau_{i,(k-1)/2}$ if k is odd.

Graded Response Model with Symmetry Constraint

$$P(Y_{pi} \geq r|\theta_p, \alpha_i, \boldsymbol{\delta}_i) = F(\theta_p - (\delta_i + \tau_{ir})),$$
$$\tau_{ij} = -\tau_{i,k+1-j}, \quad r = 1, \ldots, k.$$

Assuming that thresholds are symmetric has several consequences. First, the response of a person with latent trait $\theta_p = \delta_i$ is symmetrically distributed across response categories, that is,

$$P(Y_{pi} \geq r|\theta_p = \delta_i) = P(Y_{pi} \leq k - r|\theta_p = \delta_i), \quad r = 1, \ldots, k, \tag{4.9}$$

and therefore $P(Y_{pi} = 0|\theta_p = \delta_i) = P(Y_{pi} = k|\theta_p = \delta_i), \ldots, P(Y_{pi} = r|\theta_p = \delta_i) = P(Y_{pi} \leq k - r|\theta_p = \delta_i)$ for $r \leq 1, \ldots, k/2$. For $k = 4$ and $\theta_p = \delta_i$, it means that $P(Y_{pi} = 0) = P(Y_{pi} = 4)$, $P(Y_{pi} = 1) = P(Y_{pi} = 3)$. Figure 4.7 shows the response probabilities for the cumulative logit model with $k = 4$ in the simple case where distances between thresholds are constant. The parameter γ_i denotes the (scaled) distance between thresholds $(\tau_{i,r+1} - \tau_{i,r})/2$. It is seen that wider distances between thresholds (larger γ_i) indicate that responses in the middle categories are favored.

The symmetric distribution over response categories if $\theta_p = \delta_i$ means that a person for which the latent trait equals the location parameter is undecided. The person favors categories above the middle category as strongly as categories below the middle category although the strength of the preference depends on the distance between thresholds. The symmetry of responses at a specific value of θ_p can be seen as a symmetry postulate. It is not hard to show that postulating (4.9) is equivalent to the assumption of symmetric thresholds (4.8).

A more general consequence of symmetric thresholds is that persons with latent traits with an equal distance to the location δ_i have the same response probabilities in corresponding categories. Let the abilities for two persons p_1, p_2 be given by

$\theta_{p_1} = \delta_i + \theta, \theta_{p_2} = \delta_i - \theta$, and that means the distance between the person parameter and the item location parameter is the same for both persons but person p_1 has a high value while person p_2 has a low value. Then, the probabilities for corresponding high and low categories are the same for both persons. If k is even one obtains with $m = k/2$

$$P(Y_{p_1 i} = m + a) = P(Y_{p_2 i} = m - a), \quad a = 1, \ldots, k - m. \tag{4.10}$$

If k is odd one obtains

$$P(Y_{p_1 i} = m - 0.5 + a) = P(Y_{p_2 i} = m + 0.5 - a), \quad a = 1, \ldots, m + 0.5. \tag{4.11}$$

For $k = 4$ it means that $P(Y_{p_1 i} = 3) = P(Y_{p_2 i} = 1)$, $P(Y_{p_1 i} = 4) = P(Y_{p_2 i} = 0)$ for all θ. For $k = 3$ it means that $P(Y_{p_1 i} = 2) = P(Y_{p_2 i} = 1)$, $P(Y_{p_1 i} = 3) = P(Y_{p_2 i} = 0)$ for all θ.

Conditions (4.10) and (4.11) can be seen as general assumptions that specify in what sense responses are symmetrically distributed over response categories with a centering at δ_i.

Equidistant Thresholds

Sparser parameterizations can be obtained by assuming in addition that distances between thresholds are equal, that is, $\tau_{ir} - \tau_{i,r-1} = \tilde{\gamma}_i \geq 0$ for all r.

Then, one obtains $\tau_{ir} = \gamma_i(2r - k - 1) = \gamma_i(2(r - m) - 1), r = 1, \ldots, k$, where $\gamma_i = \tilde{\gamma}_i/2$ is the (scaled) distance between thresholds. In the latter representation, the "middle" category $m = k/2$ is used. It is the actual middle category if k is even, it is not an actual category if k is odd. In the latter case, it is in the middle between categories $0, \ldots, k/2 - 0.5$ and $k/2 + 0.5, \ldots, k$. If $k = 4$ one obtains $\tau_{i1} = -3\gamma_i, \tau_{i2} = -\gamma_i, \tau_{i3} = \gamma_i, \tau_{i4} = 3\gamma_i$. If $k = 3$ one obtains $\tau_{i1} = -2\gamma_i, \tau_{i2} = 0, \tau_{i3} = 2\gamma_i$.

> Graded Response Model with Equidistant Thresholds
>
> $$P(Y_{pi} \geq r|\theta_p, \alpha_i, \delta_i) = F(\theta_p - (\delta_i + \gamma_i(2r - k - 1))), \quad r = 1, \ldots, k.$$

From Fig. 4.8, it is seen that the modes of the response probabilities for categories $2, \ldots, k - 1$ are the same and increase if γ_i increases. For smaller values of γ_i, the response probabilities of the extreme categories 1 and k become very large, which is seen from $P(Y_{pi} = 0) = P(Y_{pi} = k) = F(\theta_p - \delta_i - \gamma_i(k - 1))$. Thus, γ_i indicates that categories in the middle are preferred, large values of γ_i are associated with small response probabilities in the extreme categories. The parameter γ_i represents the dispersion of the thresholds and also scales the person parameters. Larger γ_i means that probabilities are distinctly different from zero over a wider range of person parameters.

It is not easy to see from Fig. 4.8 how response probabilities change with the parameter γ_i at fixed values of θ_p. The general form

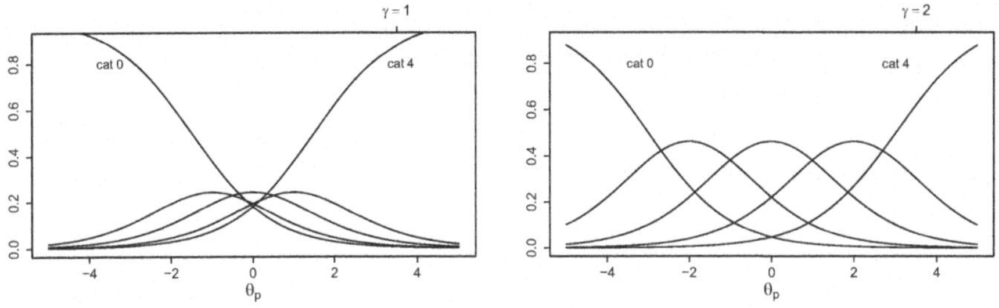

Fig. 4.8 Response probabilities for $k = 4$ and two values of γ, $(\delta_{i1}, \ldots, \delta_{ir}) = (-1.5, -0.5, 0.5, 1.5)$, $\alpha_i = 1$

$$P(Y_{pi} \geq r|\theta_p, \delta_i, \alpha_i) = F(\theta_p - \boldsymbol{\delta}_i - \gamma_i(2(r-m) - 1), \quad a = 1, \ldots, k - m$$

shows that for all $r > m$ and *all* persons the probabilities $P(Y_{pi} \geq r)$ (and the probabilities $P(Y_{pi} \leq k - r)$) decrease if γ_i increases. That means that for all persons the probabilities in extreme categories decrease with growing γ_i. For the categories in the middle $M_a = \{m - a, m - a + 1, \ldots, m + a\}$ (k even) the sum of probabilities $\sum_{r \in M_a} P(Y_{pi} = r)$ grows with γ_i for any values a and θ_p. Therefore, concerning the distribution across response categories for fixed θ_p the parameter γ_i represents the tendency to middle categories, which is not a dispersion in the classical sense. Classical dispersion parameters measure the variability with respect to a measure of central tendency.

4.3 Adjacent Categories Models and the Partial Credit Model

The graded response model contains as building blocks the (unconditional) binary models for split variables. An alternative way of obtaining an ordinal model is to use binary models conditionally. One option that is compatible with the ordering of categories is to use binary models that compare adjacent categories *given the response is in these categories*. This approach directly yields the general adjacent categories model.

General Adjacent Categories Model

$$P(Y_{pi} = r|Y_{pi} \in \{r-1, r\}, \theta_p, \alpha_i, \delta_{ir}) = F(\alpha_i(\theta_p - \delta_{ir})), \quad r = 1, \ldots, k. \tag{4.12}$$

As the graded response model, the general adjacent categories model contains three types of parameters:

- the person parameter θ_p representing ability or attitude;
- the discrimination parameters α_i, which determine the slopes of response functions;
- the location parameters collected in $\boldsymbol{\delta}_i^T = (\delta_{i1}, \ldots, \delta_{ik})$, which are on the same scale as the person parameters.

4.3.1 Partial Credit Models

The most widely used model of this class of models is the so-called partial credit model, which uses the logistic distribution function $F(.)$. Since it is the dominating model, we will focus on this model in the following. In the generalized form, it can be given by

Generalized Partial Credit Model

$$P(Y_{pi} = r|\theta_p, \alpha_i, \boldsymbol{\delta}_i) = \frac{\exp(\sum_{l=1}^{r} \alpha_i(\theta_p - \delta_{il}))}{\sum_{s=0}^{k} \exp(\sum_{l=1}^{s} \alpha_i(\theta_p - \delta_{il}))}, \quad r = 1, \ldots, k. \tag{4.13}$$

For notational convenience, the definition of the model implicitly uses $\sum_{k=1}^{0}(\theta_p - \delta_{ik}) = 0$. Again the item location parameters are collected in the vector $\boldsymbol{\delta}_i^T = (\delta_{i1}, \ldots, \delta_{ik})$.

The representation (4.13) specifies the response probabilities while the general form of adjacent categories model (4.12) specifies the conditional binary models that build the model. It is easily seen that the generalized partial credit model is an adjacent categories model since one has

$$P(Y_{pi} = r|Y_{pi} \in \{r-1, r\}, \theta_p, \alpha_i, \delta_{ir}) = \frac{\pi_{pir}}{\pi_{pi,r-1} + \pi_{pir}} =$$
$$= \frac{\exp(\alpha_i(\theta_p - \delta_{ir}))}{1 + \exp(\alpha_i(\theta_p - \delta_{ir}))} = F(\alpha_i(\theta_p - \delta_{ir})),$$

where $\pi_{pir} = P(Y_{pi} = r|\theta_p, \alpha_i, \boldsymbol{\delta}_i)$ and $F(.)$ is the logistic function. It is more difficult to derive the partial credit model in the representation (4.13) from the general model (4.12) but models are equivalent if for $F(.)$ the logistic function is chosen.

If $k = 2$ one obtains the simplest partial credit model that is not binary. The probabilities are given by

$$P(Y_{pi} = 0|\theta_p, \alpha_i, \boldsymbol{\delta}_i) = 1/\kappa_{pi},$$
$$P(Y_{pi} = 1|\theta_p, \alpha_i, \boldsymbol{\delta}_i) = \exp(\alpha_i(\theta_p - \delta_{i1}))/\kappa_{pi},$$
$$P(Y_{pi} = 2|\theta_p, \alpha_i, \boldsymbol{\delta}_i) = \exp(\alpha_i(\theta_p - \delta_{i1}) + \alpha_i(\theta_p - \delta_{i2}))/\kappa_{pi},$$

where $\kappa_{pi} = 1 + \exp(\alpha_i(\theta_p - \delta_{i1})) + \exp(\alpha_i(\theta_p - \delta_{i1}) + \alpha_i(\theta_p - \delta_{i2}))$ is a normalizing factor.

The version of the model with potentially varying discrimination parameters is known as the generalized partial credit model (GPCM) and was propagated by Muraki (1992), Muraki and Muraki (2016). The basic version of the model in which $\alpha_i = 1$ for all items is known as the partial credit model (PCM) and was propagated by Masters (1982) and Masters and Wright (1984).

Partial Credit Model

$$P(Y_{pi} = r|\theta_p, \alpha_i, \boldsymbol{\delta}_i) = \frac{\exp\left(\sum_{l=1}^{r}(\theta_p - \delta_{il})\right)}{\sum_{s=0}^{k}\exp\left(\sum_{l=1}^{s}(\theta_p - \delta_{il})\right)}, \quad r = 1, \ldots, k. \quad (4.14)$$

The partial credit model has been called a "divide-by-total" model because of the denominator in the probabilities. It is equivalent to the *polytomous Rasch model*, which is just a different parameterization, see, for example, Andrich (2010), Thissen and Steinberg (1986).

Interpretation of Parameters

There has been some confusion concerning the interpretation of parameters. In the early papers on the PCM, it has been misunderstood as a model for consecutive steps. Masters (1982) considered the example of solving the item $\sqrt{9.0/0.3 - 5} = ?$. Three levels of performance may be distinguished: no sub-problem solved (level 0), 9.0/0.3 = 30 solved (level 1), $30 - 5 = 25$ solved (level 2), $\sqrt{25} = 5$ (level 3). He described the consecutive steps in the following words: "The third step in item i, for example, is from level 2 to level 3. The difficulty of this third step governs how likely it is that a person who has already reached level 2 will complete the third step to level 3". However, the PCM does not model consecutive steps. The comparison between two adjacent categories is determined by a binary model *given* the response is in these categories, but *not* given that a certain level has been reached as in consecutive steps. Andrich (2015) even argues that the step metaphor is incompatible with the partial credit model. However, it is certainly not a model for consecutive steps. The model for consecutive steps is the sequential model, which is considered in Sect. 4.4. In later papers, the conditional structure of the PCM has been acknowledged, for example, Masters (2018) stated correctly that the local comparison is at the heart of the PCM, one conditions on a pair of adjacent categories and so eliminates all other response probabilities from consideration.

Figure 4.9 shows the response probabilities for the partial credit model with parameters $(\delta_{i1}, \ldots, \delta_{ik}) = (-2.5, -1, 2, 5)$, $k = 4$, $\alpha_i = 1$ (left), and $\alpha_i = 1.5$ (right). It is seen that the modes of the probabilities are well ordered. It is also seen that the response probabilities of adjacent categories cross at the corresponding item location parameter that means that for $\theta_p = \delta_{ir}$

$$P(Y_{pi} = r|\theta_p = \delta_{ir}, \alpha_i, \boldsymbol{\delta}_i) = P(Y_{pi} = r - 1|\theta_p = \delta_{ir}, \alpha_i, \boldsymbol{\delta}_i)$$

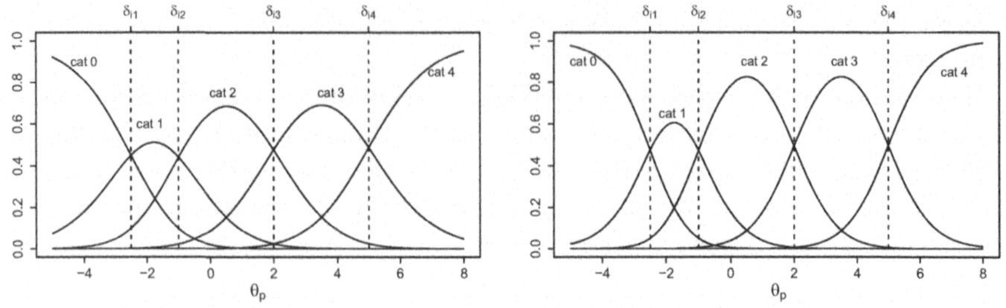

Fig. 4.9 Response probabilities for partial credit model with parameters $(\delta_{i1}, \ldots, \delta_{i4}) = (-2.5, -1, 2, 5)$, $k = 4$, left: $\alpha_i = 1$, right: $\alpha_i = 1.5$

holds, which follows immediately from the representation of the model given in Eq. (4.15). Therefore, item location parameters can be seen as the values where probabilities of adjacent categories coincide. An alternative interpretation of the location parameter is based on the conditional view, which utilizes that a binary model holds given the response is in adjacent categories. If the PCM with $\alpha_i = 1$ holds, the location parameter can be seen as the (local) difficulty that is involved when comparing adjacent categories. In the generalized PCM with varying α_is, it corresponds to the location parameter of the local binary model but does not represent the difficulty, since the difficulty is an effect of location and discrimination parameters.

Ordering of Location Parameters

An interesting case that raised some confusion is the case where item parameters are not ordered in the same way as categories. Figure 4.10 shows an example of probability functions that result if for the location parameters the ordering $\delta_{i1} \leq \cdots \leq \delta_{ik}$ does not hold. In the shown example $\delta_{i3} > \delta_{i4}$ is assumed. The result is that the probabilities of category 3 are very small. In general, probabilities tend to be very small if the corresponding location parameters do not have the same order as the categories.

It is to be emphasized that no ordering of the location parameters is needed to have a sensible model. Andrich (2013) and others claimed that location parameters have to be ordered and postulated "that the empirical ordering of categories is equivalent to the empirical ordering of successive thresholds that define adjacent categories on a continuum" (Andrich (2013), p 88). His arguments are based on the link between the Guttman space and the so-called resolved design; however, the resolved design is a complicated and unneeded concept. For a discussion of the problem of ordered categories and the conclusion that ordering is not needed, see **?** and Tutz (2020b). The latter investigates the problem by addressing Andrich's concepts explicitly.

Alternative Representations of the Generalized Partial Credit Model

The defining property of the partial credit model is seen if one considers adjacent categories. For the comparison of two adjacent categories, one obtains immediately

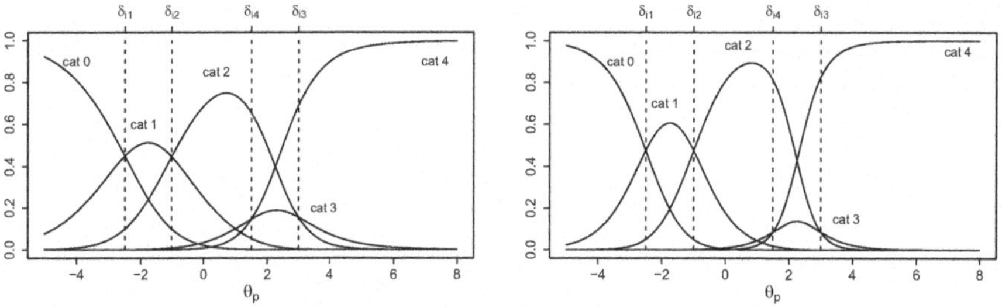

Fig. 4.10 Response probabilities for partial credit model with parameters $(\delta_{i1}, \ldots, \delta_{i4}) = (-2.5, -1, 3, 1.5)$, $k = 4$, left: $\alpha_i = 1$, right: $\alpha_i = 1.5$

$$\log\left(\frac{P(Y_{pi} = r|\theta_p, \alpha_i, \boldsymbol{\delta}_i)}{P(Y_{pi} = r-1|\theta_p, \alpha_i, \boldsymbol{\delta}_i)}\right) = \alpha_i(\theta_p - \delta_{ir}), \quad r = 1, \ldots, k. \tag{4.15}$$

That means the PCM directly compares two adjacent categories, and θ_p determines the strength of the preference for the higher category. Sometimes it is more useful to consider the representation

$$\log\left(\frac{P(Y_{pi} = r)}{P(Y_{pi} = r-1) + P(Y_{pi} = r)}\right) = \alpha_i(\theta_p - \delta_{ir}), \quad r = 1, \ldots, k, \tag{4.16}$$

which shows that the model is *locally (given response categories $r-1$, r)* a 2PL model, which reduces to the binary Rasch model with person parameter θ_p and item difficulty δ_{ir} if $\alpha_i = 1$.

An alternative form of the response probabilities in the generalized partial credit model is given by

$$P(Y_{pi} = r|\theta_p, \alpha_i, \boldsymbol{\delta}_i) = \frac{\exp(\alpha_i(r\theta_p - \sum_{l=1}^{r} \delta_{il}))}{\sum_{s=0}^{k} \exp(\sum_{l=1}^{s}(\theta_p - \delta_{il}))}. \tag{4.17}$$

A further alternative of the PCM uses a reparameterization. Let the parameters in the partial credit model ($\alpha_i = 1$) be given by

$$\delta_{il} = \delta_i + \tau_{il}, \quad l = 1, \ldots, k, \tag{4.18}$$

where $\sum_{l=1}^{k} \tau_{il} = 0$. Then one obtains

$$P(Y_{pi} = r) = \frac{1}{\gamma_{pi}} \exp(\sum_{l=1}^{r}\{\theta_p - \delta_i - \tau_{il}\}) = \frac{1}{\gamma_{pi}} \exp(r(\theta_p - \delta_i) - \sum_{l=1}^{r} \tau_{il}) \quad r = 1, \ldots, k,$$

where $\gamma_{pi} = \sum_{s=0}^{k} \exp(\sum_{l=1}^{s} \theta_p - \delta_i - \tau_{il})$ is a normalizing factor. The representation reminds of the polytomous Rasch model, which will be considered in more detail in Sect. 4.6.

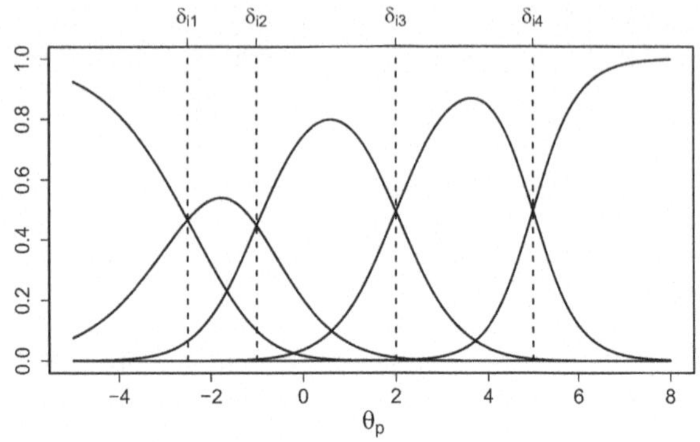

Fig. 4.11 Response probabilities for partial credit model with parameters $(\delta_{i1}, \ldots, \delta_{i4}) = (-2.5, -1, 2, 5)$, $k = 4$, $\boldsymbol{\alpha}_i = (1, 1.3, 1.5, 2)$

The parameter δ_i has several properties that make it interpretable. It represents the average of the PCM parameters

$$\delta_i = \frac{1}{k} \sum_{l=1}^{k} \delta_{il},$$

and can be considered a location parameter. Moreover, if $\theta_p = \delta_i$ one obtains $P(Y_{pi} = 0) = P(Y_{pi} = k)$. Thus δ_i is the point at which the probabilities of extreme categories are identical.

Category-Specific Discrimination Parameters

The GPCM assumes that the discrimination parameter can vary across items. They can also vary across categories, which yields the more general form of the GPCM

$$P(Y_{pi} = r|\theta_p, \boldsymbol{\alpha}_i, \boldsymbol{\delta}_i) = \frac{\exp(\sum_{l=1}^{r} \alpha_{il}(\theta_p - \delta_{il}))}{\sum_{s=0}^{k} \exp(\sum_{l=1}^{s} \alpha_{il}(\theta_p - \delta_{il}))}, \quad r = 1, \ldots, k, \quad (4.19)$$

where $\boldsymbol{\alpha}_i^T = (\boldsymbol{\alpha}_{i1}, \ldots, \boldsymbol{\alpha}_{ik})$ collects the discrimination parameters. Sometimes this generalization of the PCM is referred to as the generalized PCM, for example, in the R package tam (Robitzsch et al., 2024). Figure 4.11 shows an example of the resulting response probabilities.

Separability of Parameters and Sufficient Statistics

The simple partial credit model with $\alpha_i = 1$ inherits from the Rasch model that the comparison of items does not depend on the person parameters and the comparison of persons does not depend on the item parameters. One obtains for the comparison of items i and j

$$\log\left(\frac{P(Y_{pi} = r)/P(Y_{pi} = r - 1)}{P(Y_{pj} = r)/P(Y_{pj} = r - 1)}\right) = -(\delta_{ir} - \delta_{jr}),$$

which does not depend on the person parameter θ_p. Thus, items i and j can be compared in terms of odds ratios with the odds defined for adjacent categories without the necessity to refer to specific persons. In this sense, it is in accordance with the general principle of specific objectivity that the result of any comparison of two "objects" (items) is independent of the choice of the "agent" (person).

In the same way, the comparison between person p and person $\tilde{p}$ is obtained by

$$\log\left(\frac{P(Y_{pi}=r)/P(Y_{pi}=r-1)}{P(X_{\tilde{p}i}=r)/P(X_{\tilde{p}i}=r-1)}\right)=\theta_p-\theta_{\tilde{p}},$$

which does not depend on the item parameters. Thus, the persons p and $\tilde{p}$ can be compared without reference to the item that is used in the measurement.

Separability of parameters is a strong property that separates the measurement from the tool that is used. However, the formulae indicate theoretical separability not empirical separability. They do not show how estimates can be made independent of the other group of parameters. For the simple partial credit model with the same number of response categories in all items, this can be obtained by conditioning on the total score $Y_{p+}=r$. It can be derived that the conditional response probabilities are given by

$$P(Y_{p1}=y_{p1},\ldots,Y_{pI}=y_{pI}|\sum_{i=1}^{I}Y_{pi}=r)=\frac{\exp(-\sum_{i=1}^{I}\sum_{l=1}^{y_{pi}}\delta_{il})}{\gamma_r(\boldsymbol{\delta})},$$

where $\gamma_r(\boldsymbol{\delta})$ is a function of item parameters only. Therefore, the conditional probability does not depend on the person parameter and the corresponding conditional likelihood is a function of the item parameters only (see Sect. 4.8). The reason is that the total score Y_{p+} is a sufficient statistic for the person parameter, that is, a statistic that contains all the information on this parameter. Conditioning on the statistic makes the likelihood independent of the person parameters. The fact that the total score Y_{p+} contains all the information on the person parameter is an advantage since it is not needed to know which specific response patterns have occurred, just the sum is needed. It makes the model a member of the Rasch family, which is characterized by the existence of sufficient statistics.

The General Adjacent Categories Model*

The family of adjacent categories models is much larger because in (4.12) any strictly monotone distribution function can be used, for example, the use of the normal distribution yields a probit version of the adjacent categories model. This family of models defined in Eq. (4.12) is referred to as general adjacent categories models, of which the partial credit model is just the most prominent member.

The probabilities of the general adjacent categories model are given by

$$P(Y_{pi}=r|\theta_p,\alpha_i,\boldsymbol{\delta}_i)=\frac{\prod_{l=1}^{r}\zeta_l}{\sum_{s=0}^{k}\prod_{l=1}^{s}\zeta_l},\tag{4.20}$$

where $\prod_{l=1}^{0} \zeta_l = 1$ and $\zeta_l = F(\alpha_i(\theta_p - \delta_{il})/(1 - F(\alpha_i(\theta_p - \delta_{il}))$. Of course, (4.20) can also be understood as a definition of the general adjacent categories model. If $F(.)$ is the logistic function one obtains the (generalized) partial credit model.

For $k = 3$ the probabilities have the form

$$P(Y_{pi} = 0|\theta_p, \alpha_i, \boldsymbol{\delta}_i) = 1/\kappa_{pi}, \quad P(Y_{pi} = 1|\theta_p, \alpha_i, \boldsymbol{\delta}_i) = \zeta_1/\kappa_{pi},$$
$$P(Y_{pi} = 2|\theta_p, \alpha_i, \boldsymbol{\delta}_i) = \zeta_1\zeta_2/\kappa_{pi}, \quad P(Y_{pi} = 3|\theta_p, \alpha_i, \boldsymbol{\delta}_i) = \zeta_1\zeta_2\zeta_3/\kappa_{pi},$$

where $\kappa_{pi} = 1 + \zeta_1 + \zeta_1\zeta_2 + \zeta_1\zeta_2\zeta_3$.

In the logistic version, sufficient statistics for item and person parameters are available. While the existence of sufficient statistics is an advantage if one wants to estimate parameters conditionally it is of lesser importance if one uses marginal estimates. Then goodness-of-fit might be more important than sufficient statistics.

Example 4.2 FIMS study
We consider again the FIMS dataset from Example 4.1. Table 4.5 shows the estimated item parameters for the partial credit model (marginal estimates using ltm (Rizopoulos, 2006). It is seen that item parameters are not ordered. The reversal of item parameter ordering reflects that some categories are hardly observed. Postulating ordered thresholds would yield a worse fit since the low probability of some categories could not be represented properly. Figure 4.12 shows the corresponding response probabilities for two items. It is seen that probabilities of categories 2 and 4 are very small for all person parameters. Estimation of the generalized PCM did not converge when using function gpcm from package ltm. Fitting with program package *ltm* is obtained by the following:

```
library("ltm")
datraw<-readRDS("datAust200poly.RDS")
### reduced item set
itemsel<-c(2,3,4,6,7,8)
grrespselGEN<- grm(datraw[,itemsel], constrained = FALSE,
  IRT.param = TRUE, Hessian = FALSE,start.val = NULL,
  na.action = NULL, control = list())
deltgradGEN<-as.data.frame(grrespselGEN$coefficients)
round(t(deltgradGEN),digits=3)
```

Table 4.5 Estimated item parameters for PCM (FIMS study)

Item	δ_{i1}	δ_{i2}	δ_{i3}	δ_{i4}	δ_{i5}
2	−2.080	−0.347	−2.824	3.493	−1.267
3	−5.373	2.947	0.875	0.315	−0.169
4	−2.072	−3.132	3.274	−4.764	0.986
6	−1.514	−0.567	−2.546	2.613	0.809
7	−1.978	1.353	−1.616	2.627	−0.836
8	−2.516	−0.458	2.344	0.869	1.673

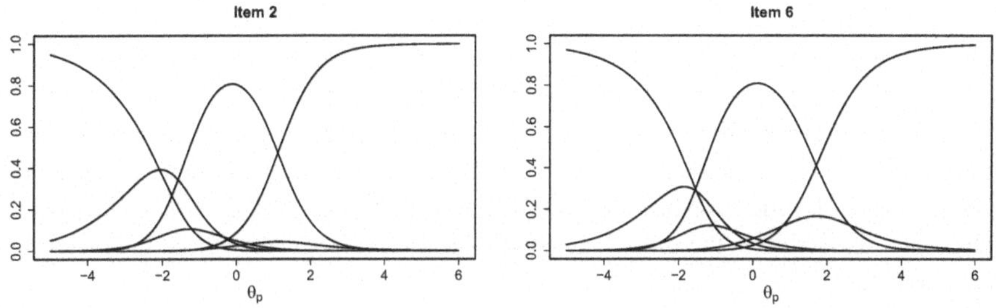

Fig. 4.12 Estimated response probabilities for some items (PCM, six items fitted)

4.3.2 *Sparser Parameterizations: The Rasch Rating Scale Model**

If one assumes that in the partial credit model the location parameter can be reparameterized by

$$\delta_{ir} = \delta_i + \tau_r, \quad r = 1, \ldots, k, \tag{4.21}$$

one obtains what Andrich (2016) calls the *Rasch Rating Scale Model* (RSM).

Rasch Rating Scale Model

$$P(Y_{pi} = r) = \frac{1}{\gamma_{pi}} \exp\left(\sum_{l=1}^{r} \{\theta_p - \delta_i - \tau_l\}\right) = \quad r, 1, \ldots, k, \tag{4.22}$$

where $\gamma_{pi} = \sum_{s=0}^{k} \exp\left(\sum_{l=1}^{s} \theta_p - \delta_i - \tau_l\right)$ is a normalizing factor.

To obtain identifiability, the τ-parameters have to be constrained, a commonly used constraint is $\sum_{l=1}^{k} \tau_l = 0$. The parameter θ_p contains the location of the person on the latent continuum, the parameter δ_i is the location of the item, and the τ-parameters are considered as thresholds.

In contrast to the PCM in the RSM, each item has just one location parameter δ_i. Using $\delta_{ir} = \delta_i + \tau_r$ and $\sum_{l=1}^{k} \tau_l = 0$ immediately yields $\delta_i = \sum_{l=1}^{k} \delta_{ir}/k$. Thus δ_i

represents the average over thresholds. The specific parameterization of thresholds in (4.21) also means that the thresholds do not vary over items, which is a rather strong assumption.

Alternative representations are

$$P(Y_{pi} = r) = \frac{1}{\gamma_{pi}} \exp\left(r(\theta_p - \delta_i) - \sum_{l=1}^{r} \tau_l \right) \quad r = 1, \ldots, k,$$

and

$$\log \left(\frac{P(Y_{pi} = r)}{P(Y_{pi} = r - 1)} \right) = \theta_p - \delta_i - \tau_r = \theta_p - (\delta_i + \tau_r), \quad r = 1, \ldots, k. \quad (4.23)$$

The local binary Rasch model given in (4.23), which distinguishes between categories r and $r - 1$, given one of the two categories is chosen, uses the item difficulty $\delta_i + \tau_l$. Typically, the thresholds τ_r are ordered, that is, $\tau_1 \leq \tau_2 \cdots \leq \tau_k$, however, there is no restriction in the model, although Andrich (2016) argues that the ordering is needed.

Equal Distance Thresholds Model

Andrich (1982) proposed a further restricted model that reduces the item parameters to a location and a scale parameter. If one assumes equal distances between thresholds, more concrete, $\tau_r - \tau_{r-1} = \tilde{\gamma}$, one obtains with $\sum_{l=1}^{k} \tau_l = 0$ that $\sum_{l=1}^{r} \tau_l = \tilde{\gamma} r(r - k)/2$. The corresponding model, which will be referred to as *Equal Distance Rating Scale Model* (EDRSM), is given by

$$P(Y_{pi} = r) = \frac{1}{\gamma_{pi}} \exp(r(\theta_p - \delta_i) + \gamma r(k - r)) \quad r = 1, \ldots, k,$$

where $\gamma = \tilde{\gamma}/2$ is the (scaled) distance between thresholds. Alternatively, one has

$$\log \left(\frac{P(Y_{pi} = r)}{P(Y_{pi} = r - 1)} \right) = \theta_p - \delta_i + \gamma(k - 2r + 1), \quad r = 1, \ldots, k,$$

which uses that the τ-parameters are given by $\tau_r = \gamma(2r - k - 1)$.

The parameter γ reflects the variability of the τ-parameters, however in a specific way. The constraint $\sum_{l=1}^{k} \tau_l = 0$ makes the thresholds *symmetric*, that is, $\tau_r = -\tau_{k+1-r}$. For example, if $k = 4$ one obtains

τ_1	τ_2	τ_3	τ_4
-3γ	$-\gamma$	γ	3γ.

For *positive* γ, the thresholds are ordered such that $\tau_1 \leq \tau_2 \leq \cdots \leq \tau_k$ and the variability of the τ-parameters increases with increasing γ. In particular, one obtains for the distance between the extreme thresholds $\tau_k - \tau_1 = 2(k - 1)\gamma$. The widening

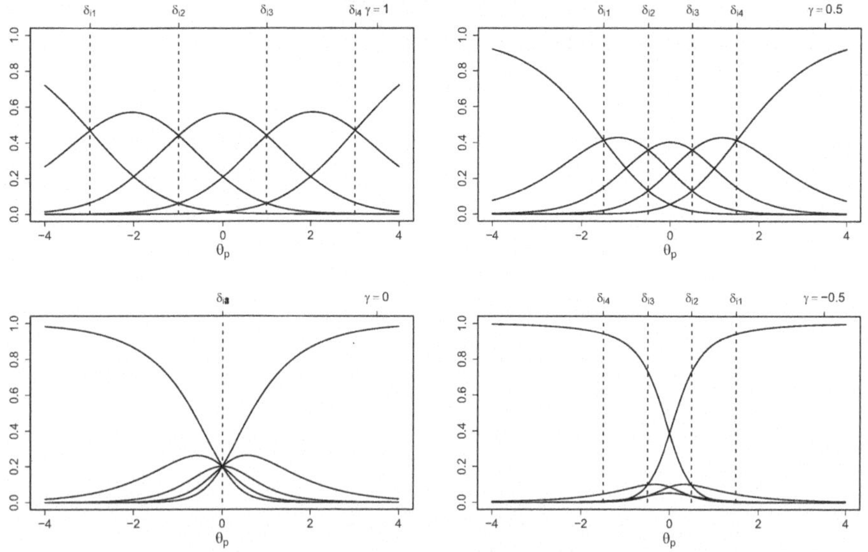

Fig. 4.13 Response probabilities for Rasch rating scale model with equidistant thresholds model for several values of γ_i and $k = 4$

of thresholds when γ is increased means that the probabilities in middle categories become larger (see Fig. 4.13). Since thresholds are local when distinguishing between two adjacent categories given the response is in these categories there is no need to postulate that parameters δ_{ir} are ordered. Thus, the (scaled) distance between thresholds γ can also be negative, which means that the order of thresholds is reversed, $\tau_1 \geq \tau_2 \geq \cdots \geq \tau_k$. Negative values indeed yield proper distributions (see Fig. 4.13). If γ is negative the probabilities in middle categories become very small.

Andrich (1982) refers to γ as a scale parameter but also as an index of dispersion. The interpretation as a dispersion parameter has to be seen with caution since decreasing γ means larger distances between extreme thresholds if γ is negative, thus smaller values of γ indicate larger dispersion of τ-parameters. The parameter γ indicates the tendency to middle categories rather than dispersion. As is seen from Fig. 4.13, the probabilities in middle categories become larger if γ increases. This becomes obvious by considering

$$\log\left(\frac{P(Y_{pi} = r)}{P(Y_{pi} = r - 1)}\right) = \theta_p - \delta_i - \gamma(2(r - m) - 1), \quad r = 1, \ldots, k,$$

which uses that the τ-parameters are given by $\tau_r = \gamma(2r - k - 1) = \gamma(2(r - m) - 1)$, where $m = k/2$ denotes the middle category. It shows that the probabilities for the larger category r as compared to the previous category $r - 1$ decrease if $r > m$ and increase if $r < m$.

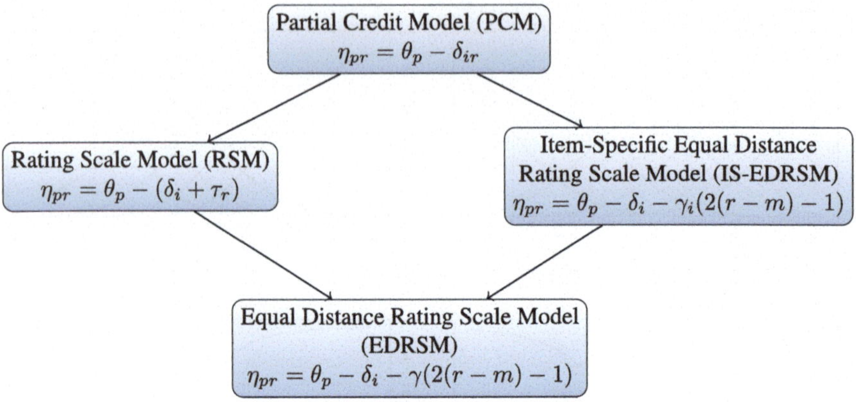

Fig. 4.14 Hierarchy of adjacent categories models

The equal distance rating scale model is even more restrictive than the rating scale model. In a more flexible model at least the parameter γ can depend on the item yielding the *Item-Specific Equal Distance Rating Scale Model* (IS-EDRSM).

Item-Specific Equal Distance Rating Scale Model

$$\log\left(\frac{P(Y_{pi}=r)}{P(Y_{pi}=r-1)}\right) = \theta_p - \delta_i - \gamma_i(2(r-m)-1), \quad r = 1,\dots,k. \tag{4.24}$$

It can be derived from the partial credit model by using the reparameterization $\delta_{ir} = \delta_i + \tau_{ir}$ with the assumptions $\sum_{l=1}^{k} \tau_{il} = 0$, $\tau_{ik} - \tau_{i,k-1} = 2\gamma_i$. In the model, the items are characterized by just two parameters, the location parameter δ_i and the tendency to the middle parameter γ_i.

Figure 4.14 shows the resulting hierarchy of models with η_{pr} denoting the linear term in the representation $\log(P(Y_{pi}=r)/P(Y_{pi}=r-1)) = \eta_{pr}$. While the Equal Distance Rating Scale Model might be too simple to fit the data well, the Item-Specific Equal Distance Rating Scale Model offers an alternative, which has a sparser parameterization than the partial credit model but not as sparse as the Equal Distance Rating Scale Model. Table 4.6 gives an overview on the various versions of adjacent categories models.

4.4 Sequential Models

In achievement tests frequently items are used that are solved in consecutive observed steps. For example, a mathematical problem may have the form: $(\sqrt{49} - 9)^3 = ?$. One can distinguish four levels: no problem solved (level 0), $\sqrt{49} = 7$ solved (level

Table 4.6 Adjacent categories models

Model structure	$P(Y_{pi} = r \| Y_{pi} \in \{r-1, r\}) = F(\eta_{pir})$	
	Restriction	Predictor
General adjacent categories model	–	$\eta_{pir} = \alpha_i(\theta_p - \delta_{ir})$
Generalized PCM	$F(.)$ logistic	$\eta_{pir} = \alpha_i(\theta_p - \delta_{ir})$
PCM	$F(.)$ logistic, $\alpha_i = 1$	$\eta_{pir} = \theta_p - \delta_{ir}$
PCM, polytomous Rasch representation	$F(.)$ logistic, $\alpha_i = 1$	$\eta_{pir} = \theta_p - \delta_i + \tau_{ir}$
Rasch rating scale model	$F(.)$ logistic, $\alpha_i = 1$, $\tau_{ir} = \tau_r$	$\eta_{pir} = \theta_p - \delta_i + \tau_r$

1), $7 - 9 = -2$ solved (level 2), $(-2)^3 = -8$ solved (level 3). Obviously, the sub-problems have to be solved in a consecutive way. A sub-problem can only be solved if the all the previous sub-problems have been solved. A model that explicitly models the solving of sub-problems is the sequential model.

In each of the consecutive steps, a binary response model is used. Let $Y_{pi}^{(r)}$ denote the binary response in the rth step. In the first step, the respondent tries to master the transition from level 0 to level 1. Let the corresponding model be

$$P(Y_{pi}^{(1)} = 1) = F(\alpha_{i1}(\theta_p - \delta_{i1})).$$

If the person does not master the first step, the process stops and one observes $Y_{pi} = 0$. Otherwise, $Y_{pi} \geq 1$ and the person tries to solve the next step. In general, the rth step, which is only tried if the previous steps were successful, is determined by

$$P(Y_{pi}^{(r)} = 1 | Y_{pi} \geq r - 1) = F(\alpha_{ir}(\theta_p - \delta_{ir})), \quad r = 1, \ldots, k.$$

If the person does not master the step, the process stops and one observes $Y_{pi} = r - 1$. Otherwise, $Y_{pi} \geq r$ and the person tries to solve the next step. The process is illustrated in Fig. 4.15 for response categories $\{0, 1, 2, 3\}$. It is straightforward to derive the (generalized) sequential model.

Generalized Sequential Model

$$P(Y_{pi} \geq r | Y_{pi} \geq r - 1, \theta_p, \alpha_i, \boldsymbol{\delta}_i) = F(\alpha_{ir}(\theta_p - \delta_{ir})),$$
$$r = 1, \ldots, k. \quad (4.25)$$

The important feature of the model is that the transition from one level to the next is modeled *conditionally* (given the previous level has been reached). For each conditional step, the binary model could also be motivated by latent random variables

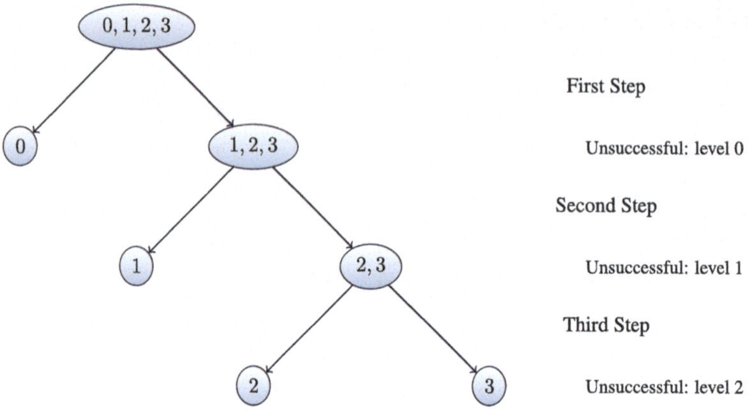

Fig. 4.15 The sequential model as a hierarchically structured model

$\tilde{Y}_{pi}^{(r)} = \theta_p + \varepsilon_{pi}$, where ε_{pi} is a noise variable with symmetric continuous distribution function $F(.)$. Given the rth step is reached one has $Y_{pi}^{(r)} = 1$ or, equivalently, $Y_{pi} \geq r$, if $\tilde{Y}_{pi}^{(r)} > \delta_{ir}$, and $Y_{pi}^{(r)} = 0\,(Y_{pi} = r)$ otherwise. Thus, in proficiency tests, it is straightforward to interpret the person parameter as the ability and δ_{ir} as a threshold that determines the difficulty of the rth step. Subsequent steps can be easier than previous steps, thus item difficulties are not necessarily ordered. In the mathematical problem $(\sqrt{49} - 9)^3 =?$ step 2 $(7 - 9)$ is certainly easier to master than step 1 $(\sqrt{49} = 7)$. However, sub-problem 2 can be only solved after step 1 was successful. Therefore, the item parameters have *local meaning*, they refer to the difficulty in a step given that all previous steps have been mastered. By contrast, the same ability parameter is present in each of the steps, which makes the model unidimensional in terms of person parameters. If discrimination parameters vary the transition to higher categories is determined by a three-parameter model, in which α_{ir} *and* δ_{ir} determine the difficulty.

The binary models that are contained in the sequential models are conditional; they consider if a response larger or equal r is observed *given* $Y_{pi} \geq r - 1$. In the rth step, one compares the groups $\{r - 1\}$ and $\{r, \ldots, k\}$. The response is in the latter set if the transition is successful. It is noteworthy that the binary variables that are used when modeling steps are equivalent to the split variables, and the model can also be formulated in terms of split variables:

$$P(Y_{pi}^{(r)} = 1 | Y_{pi}^{(r-1)} = 1, \ldots, Y_{pi}^{(1)} = 1) = F(\alpha_{ir}(\theta_p - \delta_{ir})), \quad r = 1, \ldots, k. \tag{4.26}$$

The condition $Y_{pi}^{(r-1)} = 1, \ldots, Y_{pi}^{(1)} = 1$ means that all previous steps were successful. The sequence of split variables $(Y_{pi}^{(1)} \ldots, Y_{pi}^{(k)})$ again forms a Guttman space, in which a sequence of ones is followed by a sequence of zeros.

The probabilities of the sequential model are obtained by

$$P(Y_{pi}=r)=(1-F(\alpha_{i,r+1}(\theta_p-\delta_{i,r+1})))\prod_{j=1}^{r}F(\alpha_{ij}(\theta_p-\delta_{ij})),\quad r=0,\ldots,k,$$

where $\prod_{j=1}^{0}\equiv 1$ and $\delta_{i,k+1}=\infty$. For $k=2$ they are given by

$$P(Y_{pi}=0)=1-F(\alpha_{i1}(\theta_p-\delta_{i1})),$$
$$P(Y_{pi}=1)=(1-F(\alpha_{i2}(\theta_p-\delta_{i2})))F(\alpha_{i1}(\theta_p-\delta_{i1})),$$
$$P(Y_{pi}=2)=F(\alpha_{i1}(\theta_p-\delta_{i1}))F(\alpha_{i2}(\theta_p-\delta_{i2})).$$

The probabilities reflect the sequential nature of r successful steps with probabilities $F(\theta_p-\delta_{ij})$ followed by one unsuccessful step, which is the step to reach category $r+1$. In this step, the probability of failing is $1-F(\theta_p-\delta_{i,r+1})$.

The logistic version of the model, in which $F(.)$ is chosen as the logistic distribution function, is also called logistic sequential model. It can be given in the alternative form of a *continuation ratio model*,

$$\log\left(\frac{P(Y_{pi}\geq r)}{P(Y_{pi}=r-1)}\right)=\alpha_{ir}(\theta_p-\delta_{ir}),\quad r=1,\ldots,k, \tag{4.27}$$

(**?**). The logits on the left-hand side directly compares the probability of a response in the categories $\{r,\ldots,k\}$ to the probability of a response in category $\{r-1\}$. A disadvantage of the model representation (4.27) is that it does not directly show the underlying process. The implicit conditioning on responses $Y_{pi}\geq r$, which is essential for the interpretation of the model parameters, tends to get lost.

Figure 4.16 shows the probabilities for the logistic sequential model with parameters $\alpha_{ir}=1$, $(\delta_{i1},\ldots,\delta_{ir})=(-3,-1,1,3.5)$ on the left. The right panel shows the probability if the second threshold is harder to overcome, that is, $\delta_{i2}=-1$ is replaced by $\delta_{i2}=0$. Consequently, for abilities in the middle range, the probabilities for category 1 are larger while the probability categories 2, 3, 4 are smaller in the right panel.

Figure 4.17 shows the probabilities for the logistic sequential model with essentially the same parameters as Fig. 4.16, only the discrimination parameter has been changed to $\alpha_{ir}=2$. It is seen that response probabilities become more distinct. Categories are more clearly separated.

The sequential model has been considered in a simpler version (with $\alpha_{ir}=1$) by Tutz (1990) and under the name *step model* by Verhelst et al. (1997). A rating scale version of the model, in which the parameter δ_{ir} is split up into a an item location parameter δ_i and a step parameter τ_r with $\sum_r \tau_r=0$, is found in Tutz (1990), extended versions with predictor $\alpha_{ir}(\theta_p-\delta_{ir})$ and nonparametric versions have been considered by Hemker et al. (2001). Deng and Bolt (2016) considered a sequential response model for multiple-choice items. Sequential models are special cases of tree-based models, which are considered in more detail in Chap. 9.

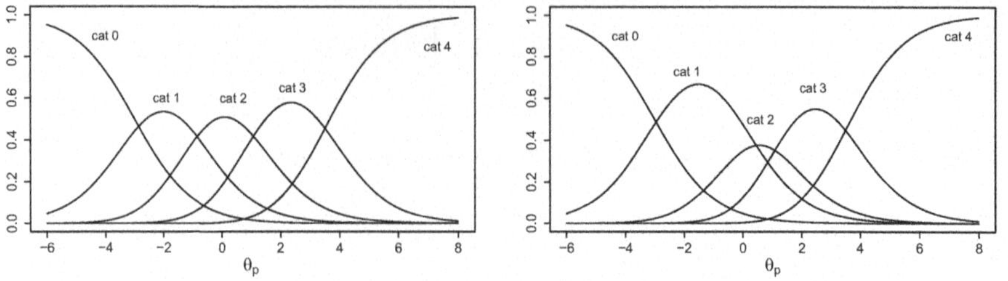

Fig. 4.16 Response probabilities for logistic sequential model with parameters $\alpha_{ir} = 1$, $(\delta_{i1}, \ldots, \delta_{ir}) = (-3, -1, 1, 3.5)$ (left) and parameters $(\delta_{i1}, \ldots, \delta_{ir}) = (-3, 0, 1, 3.5)$ (right), left and right differ only in δ_{i2}

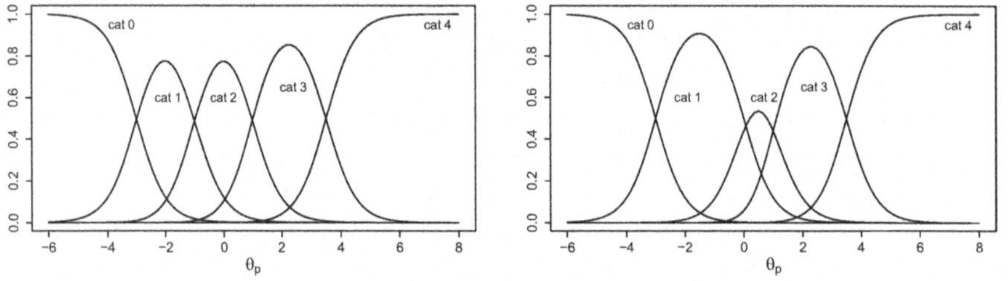

Fig. 4.17 Response probabilities for logistic sequential model with parameters $\alpha_{ir} = 2$, $(\delta_{i1}, \ldots, \delta_{ir}) = (-3, -1, 1, 3.5)$ (left) and parameters $(\delta_{i1}, \ldots, \delta_{ir}) = (-3, 0, 1, 3.5)$ (right), left and right differ only in δ_{i2}

Information Function

The information function can be computed by using (4.7) with response probabilities given by

$$\pi_{ir}(\theta_p) = (1 - F(\alpha_{i,r+1}(\theta_p - \delta_{i,r+1}))) \prod_{j=1}^{r} F(\alpha_{ij}(\theta_p - \delta_{ij})), \quad r = 0, \ldots, k,$$

and derivatives given by

$$\pi'_{ir}(\theta_p) = \frac{\partial \pi_{ir}}{\partial \theta_p} = e^{\log(\pi_{ir})}\Big(\frac{-\alpha_{i,r+1} F'(\alpha_{i,r+1}(\theta_p - \delta_{i,r+1}))}{1 - F(\alpha_{i,r+1}(\theta_p - \delta_{i,r+1}))} + \sum_{j=1}^{r} \frac{\alpha_{ij} F'(\alpha_{ij}(\theta_p - \delta_{ij})}{F(\alpha_{ij}(\theta_p - \delta_{ij})}\Big),$$

where $F'(\eta) = \frac{\partial F(\eta)}{\partial \theta_p}$. For the logistic function, $F'(\eta)$ is given by $F'(\eta) = F(\eta)(1 - F(\eta))$.

The functions corresponding to the probabilities given in Fig. 4.16 are shown in Fig. 4.18. Similar as in cumulative models they are not necessarily concave. Figure 4.19 shows the information function if $\alpha_{ir} = 2$, which yields more wiggly functions.

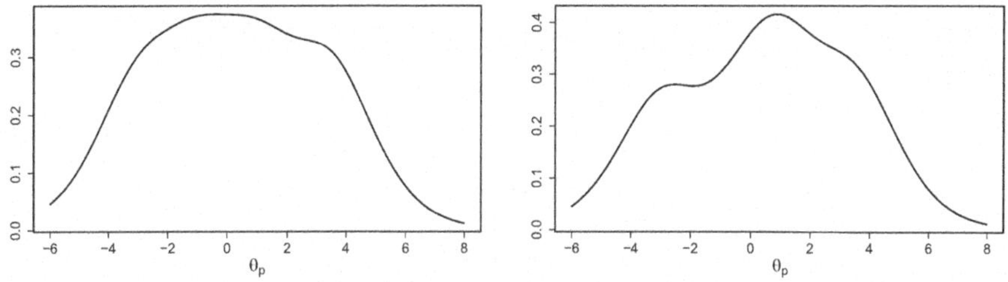

Fig. 4.18 Information for logistic sequential model with parameters $\alpha_{ir} = 1$, $(\delta_{i1}, \ldots, \delta_{ir}) = (-3, -1, 1, 3.5)$ (left) and parameters $(\delta_{i1}, \ldots, \delta_{ir}) = (-3, 0, 1, 3.5)$ (right), left and right differ only in δ_{i2}

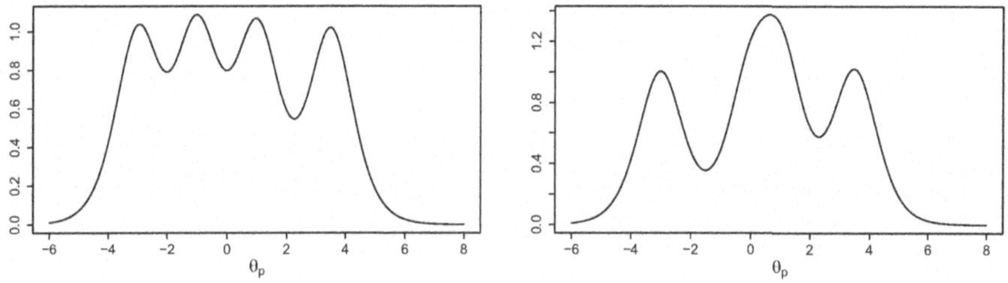

Fig. 4.19 Information for logistic sequential model with parameters $\alpha_{ir} = 2$, $(\delta_{i1}, \ldots, \delta_{ir}) = (-3, -1, 1, 3.5)$ (left) and parameters $(\delta_{i1}, \ldots, \delta_{ir}) = (-3, 0, 1, 3.5)$ (right), left and right differ only in δ_{i2}

Separability of Parameters

For specific sequential models parameters may be separated. Let $F(.)$ be the logistic function and $\alpha_{ir} = 1$ for all items and categories. Then one obtains for the *continuation odds ratio*

$$\gamma_{pi} = \frac{P(Y_{pi} \geq r|\theta_p, \delta_i)}{P(Y_{pi} = r-1|\theta_p, \delta_i)} = \exp(\theta_p - \delta_{ir}).$$

It means that for two persons p_1, p_2

$$\frac{\gamma_{p_1 i}}{\gamma_{p_2 i}} = \frac{\exp(\theta_{p_1})}{\exp(\theta_{p_2})}$$

holds, which does not depend on the item. Comparisons of persons can be carried out independent of the item involved. In the same way, the continuation odds ratio for two items does not depend on the person.

Example 4.3 FIMS study

We consider again the FIMS dataset from Example 4.2. Table 4.7 shows the estimated item parameters for the sequential logistic model. The parameters which represent the difficulties of single steps vary strongly across steps. For example, in item 2 they

Table 4.7 Estimated item parameters for sequential logistic model (six items)

Item	δ_{i1}	δ_{i2}	δ_{i3}	δ_{i4}	δ_{i5}
2	−1.521	0.061	0.012	4.536	1.348
3	−2.993	4.565	1.976	1.617	2.168
4	−2.993	−1.516	1.522	0.000	3.787
6	−0.880	0.162	0.403	4.913	3.404
7	0.322	2.359	0.943	3.784	1.664
8	−0.323	2.380	4.479	3.064	3.958

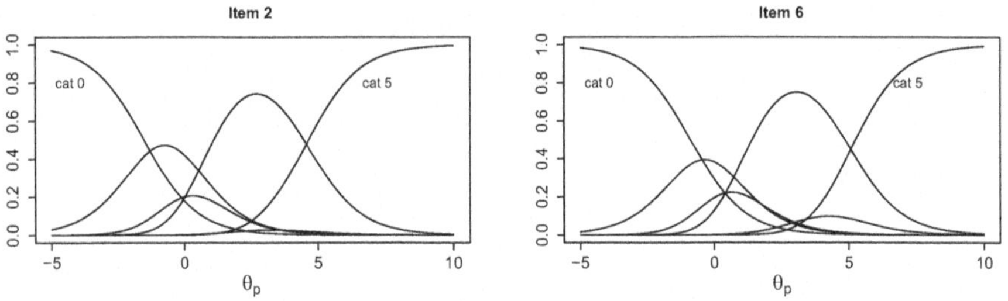

Fig. 4.20 Estimated response probabilities for some items (sequential model)

take values between −1.52 and 4.43. Let us consider the response probabilities for persons with $\theta_p = 0$. For that value the transition from level 0 to a higher level is determined by $\delta_{i1} = -1.52$, which means it is an easy step and the probability of staying in level 0 is just 0.18. The next two steps are determined by parameters close to zero, which means transition probabilities are close to 0.5. The transition to levels above or equal 4 becomes very hard ($\delta_{i1} = 4.53$). Consequently, the probabilities of category 4 and 5 are almost zero., also seen from Fig. 4.20, which shows the resulting probabilities for items 2 and 6. □

4.5 Common Structure of Ordinal Models

All the models considered so far not only contain binary models that split categories into two subsets but can also be constructed from these binary models. In the partial credit model and the sequential model, the splits are conditional whereas in the cumulative model the splits are simultaneous but not conditional.

Table 4.8 gives an overview on the models. The first column shows the logistic versions of the models, which are the simplest ones. It shows which categories or groups of categories are compared. In particular, it is seen what type of logits are determined by the difference between person parameter and item parameter, $\theta_p - \delta_{ir}$. For example, in the partial credit model, one has the adjacent categories logits

Table 4.8 Overview of traditional ordinal models

	Category representation	Conditional representation	Conditional representation
	Logistic version	General version	With split variables
	$\log(.) = \alpha_i(\theta_p - \delta_{ir})$	$P(.) = F(\alpha_i(\theta_p - \delta_{ir}))$	$P(.) = F(\alpha_i(\theta_p - \delta_{ir}))$
Cumulative	$\log\left(\frac{P(Y_{pi} \geq r)}{P(Y_{pi} < r)}\right)$	$P(Y_{pi} \geq r)$	$P(Y_{pi}^{(r)} = 1)$
Adjacent	$\log\left(\frac{P(Y_{pi} = r)}{P(Y_{pi} = r-1)}\right)$	$P(Y_{pi} = r \mid Y_{pi} \in \{r-1, r\})$	$P(Y_{pi}^{(r)} = 1 \mid Y_{pi}^{(r-1)} = 1, Y_{pi}^{(r+1)} = 0)$
Sequential	$\log\left(\frac{P(Y_{pi} \geq r)}{P(Y_{pi} = r-1)}\right)$	$P(Y_{pi} \geq r \mid Y_{pi} \geq r-1)$	$P(Y_{pi}^{(r)} = 1 \mid Y_{pi}^{(r-1)} = 1)$

$\log(P(Y_{pi} = r)/P(Y_{pi} = r-1))$, in the sequential model, one has the continuation odds ratios $\log(P(Y_{pi} \geq r)/P(Y_{pi} = r-1))$. In the middle column, the general conditional representations of the models are given. In these representations, the distribution function $F(.)$ can be any strictly monotonic distribution function. It shows which *conditional* binary response models are contained in the ordinal model. In the case of the graded response model, the condition is empty since it is a non-conditional model.

The third column shows the representation of the general models with split variables. It also shows clearly the conditioning implicitly contained in the models. It should be emphasized that in *all* the models the split variables $(Y_{pi}^{(1)} \dots, Y_{pi}^{(k)})$ form a Guttman space, that is, a space of tuples where a sequence of ones is followed by a sequence of zeros. They can be seen as generating the Guttman space, which is always defined and is not linked to any specific model.

As mentioned already, the ordinal response in all models is given by the sum of split variables, $Y_{pi} = Y_{pi}^{(1)} + \cdots + Y_{pi}^{(k)}$. This follows directly from the definition of the split variables because one has

$$Y_{pi} = r \Leftrightarrow (Y_{pi}^{(1)}, \dots, Y_{pi}^{(r)}, \dots, Y_{pi}^{(k)}) = (1, \dots, 1, 0, \dots, 0),$$

where $Y_{pi}^{(r)}$ is the last one of the sequence of split variables with a value 1. This general property makes the split variables a versatile tool that can be used to characterize the structure of models.

The hierarchy of classical ordinal models is shown in Fig. 4.21, which is an extended version of the structure already shown in Fig. 4.1. Now the parameterizations that yield special cases of the adjacent categories and the cumulative models are included.

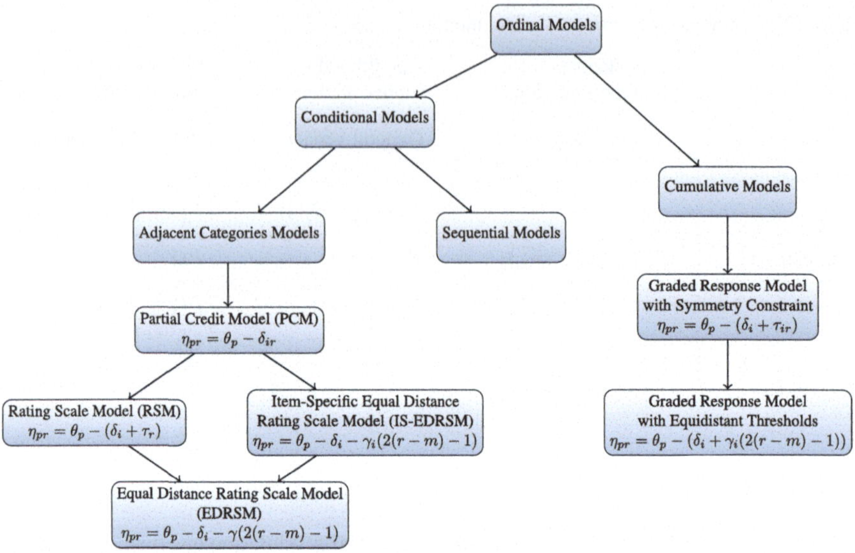

Fig. 4.21 Taxonomy of classical ordinal latent trait models with parameterizations

Interpretation of Parameters

The dichotomizations also clarify the meaning of parameters. In the graded response model, the split variables $Y_{pi}^{(r)}$, which distinguish between a response $Y_{pi} \geq r$ (strong performance) and $Y_{pi} < r$ (weak performance), are directly linked to the difference between "ability" and item parameter, $\theta_p - \delta_{ir}$. The corresponding binary models for split variables are unconditional and have to hold simultaneously. This allows to see the item difficulties as thresholds that are necessarily ordered, and which have to be exceeded to obtain higher levels. However, the variables $Y_{pi}^{(r)}$ should not be seen as steps. $Y_{pi}^{(r)} = 1$ simply denotes that a person has at least performance level r. Since performance levels are ordered, it means the performance cannot be below level r, or, in split variables, $Y_{pi}^{(1)} = \cdots = Y_{pi}^{(r)} = 1$, which is the Guttman property of the binary responses. One observes $Y_{pi} = r$, if, in addition $Y_{pi}^{(r+1)} = 0$, which means that the performance is below level $r + 1$. No steps or transitions are needed to explain the level of performance. As Andrich (2015) argues, if a performance like acting is to be classified according to some protocol, the judge places the person's performance in one of the categories on the trait, not how the person transitioned in getting to the category. Even in simple binary models for problem-solving one observes if the problem was solved or not, but not the transition. Thus, when considering ordinal models and the binary models contained in them there is no reason to construct a transition. It might be misleading and is not compatible with the underlying process, which is determined by *simultaneous* dichotomizations or the placing on the continuum of the latent scale, which is divided by the thresholds $\delta_{i1} \leq \cdots \leq \delta_{i,k}$.

Interpretation of parameters is quite different in conditional models. Let us start with the sequential model since it is by construction a step or transition model. The split variables representation, $P(Y_{pi}^{(r)} = 1|Y_{pi}^{(r-1)} = 1) = F(\theta_p - \delta_{ir})$, shows that the difference between ability and item difficulty determines whether the performance is above or in category r given at least performance level $r-1$ has been reached. It makes the parameter δ_{ir} a local threshold parameter. There is no ordering of thresholds involved since subsequent steps might be easier than previous steps. It should also be noted that the condition $Y_{pi}^{(r-1)} = 1$ is equivalent to $Y_{pi}^{(1)} = \dots = Y_{pi}^{(r-1)} = 1$ since the split variables form a Guttman space. The condition implies that all previous steps were successful.

In the partial credit model, the decision that the performance is at least in category r as a function of the difference between ability and item parameter is under the condition $Y_{pi} \in \{r-1, r\}$ or, equivalently, $Y_{pi}^{(r-1)} = 1, Y_{pi}^{(r+1)} = 0$. Thus, the binary models are conditional models and parameters should be interpreted with reference to the conditional structure. One consequence of the conditional parameterization is that thresholds do not have to be ordered even though there has been some discussion on the ordering of thresholds (**?**Andrich, 2013, 2015; Tutz, 2020b).

Separability of Parameters*

It has already been shown that comparisons of items that do not depend on persons, as well as comparisons of persons that do not depend on items, are possible for specific versions of polytomous models. In the following, a general concept of specific objectivity or separability of parameters is given. Let $\boldsymbol{\pi}_{pi}^T = (\pi_{pi0}, \dots, \pi_{pik})$ denote the vector of response probabilities for person p and item i with components given by $\pi_{pir} = P(Y_{pi} = r)$. For a (unidimensional) latent trait model, *invariance of comparison of person parameters* (*specific objectivity for the comparison of persons*) holds if a parameterization and a transformation function called comparator C_{pers} exist such that for all persons p_1, p_2

$$C_{\text{pers}}(\boldsymbol{\pi}_{p_1 i}(\theta_{p_1}, \boldsymbol{\delta}_i), \boldsymbol{\pi}_{p_2 i}(\theta_{p_2}, \boldsymbol{\delta}_i))) = \theta_{p_1} - \theta_{p_2},$$

which means, in particular, that the comparison does not depend on the item parameters. A simple example is the graded response model with $\alpha_i = 1$ for all items. For any $r \in \{1, \dots, k\}$, one has

$$C_{\text{pers}}(\boldsymbol{\pi}_{p_1 i}, \boldsymbol{\pi}_{p_2 i}) = \frac{\pi_{p_1 ir} + \dots + \pi_{p_1 ik}}{\pi_{p_1 i0} + \dots + \pi_{p_1 i,k-1}} - \frac{\pi_{p_2 ir} + \dots + \pi_{p_2 ik}}{\pi_{p_2 i0} + \dots + \pi_{p_2 i,k-1}} = \theta_{p_1} - \theta_{p_2}.$$

The comparison of items is slightly more difficult since items are determined by more than one parameter. For a latent trait model with k-dimensional item parameters $\boldsymbol{\delta}_i$ *invariance of comparison of item parameters* (*specific objectivity for the comparison of items*) holds if a parameterization and a set of transformation functions (comparators) $C_{\text{it},i}$ exist such that for two items i_1, i_2 the differences of item parameters are *uniquely* determined in the form:

$$C_{\text{it},i}(\boldsymbol{\pi}_{pi_1}(\theta_p, \boldsymbol{\delta}_{i_1}), \boldsymbol{\pi}_{pi_2}(\theta_p, \boldsymbol{\delta}_{i_2}))) = \delta_{i_1 q} - \delta_{i_2 r}.$$

In contrast to the comparison of person parameters, it is not sufficient to postulate the existence of just one comparator function. Since each item has several parameters in the general case more than one comparator function is needed to determine all differences uniquely.

For the graded response model with $\alpha_i = 1$, let comparator functions $C_{\text{it},r,q}$ with $r, q \in \{1, \dots, k\}$ be given by

$$C_{\text{it},rq}(\boldsymbol{\pi}_{pi_1}, \boldsymbol{\pi}_{pi_2}) = \log(\frac{\pi_{pi_1r} + \dots + \pi_{pi_1k}}{\pi_{pi_10} + \dots + \pi_{pi_1,r-1}}) - \log(\frac{\pi_{pi_2q} + \dots + \pi_{pi_2k}}{\pi_{pi_20} + \dots + \pi_{pi_2,q-1}}),$$
$$= \delta_{i_1q} - \delta_{i_2r},$$

which determine the differences between item parameters uniquely.

Postulating that the *difference* of person (item) parameters is independent of items (persons) seems rather restrictive. However, it is only postulated that a parameterization has to exist, for which this holds. In the simplest example, the binary Rasch model $P(Y_{pi} = 1) = \exp(\theta_p - \delta_i)/(1 + \exp(\theta_p - \delta_i))$, the log-odds ratio for two persons p_1, p_2 is given by $\log(\gamma_{p_1i}) - \log(\gamma_{p_2i}) = \theta_{p_1} - \theta_{p_2}$ (see Eq. (2.10)). Thus, there is a parameterization and a transformation for which one obtains the difference of person parameters. For alternative parameterizations, for example, $\theta_p^* = \exp(\theta_p)$, $\delta_i^* = \exp(\delta_i)$, which is also often used, the log-odds ratio for two persons p_1, p_2 is not given as the difference between person parameters. However, it suffices that there is one parameterization that fulfills the postulate. Wider concepts of specific objectivity that allow for qualitative comparisons rather than quantitative ones have been considered by Irtel (1995).

Specific objectivity in the sense used here holds for all logistic versions of the models given in the left column of Table 4.8 if discrimination parameters are fixed, that is, $\alpha_i = 1$ holds for all items. At the heart of the concept of specific objectivity as given here is that comparisons of items (persons) are possible and independent on persons (items). It means that measurement is possible without referring to the other set of parameters. It does not refer to the existence of sufficient statistics. If sufficient statistics exist, which holds for the partial credit model but not for the graded response and the sequential model, it just means that the information on specific parameters is contained in a statistic. It allows for conditional estimation methods but that does not mean that estimation does not depend on the parameters for which sufficient statistics are available. For general concepts of the invariance of comparisons, see also Tutz (2024a).

Collapsing of Categories

Consideration of the binary models contained in ordinal models also explains why some models are robust against the collapsing of categories. In general, if a model holds for the original categories it does not necessarily hold if adjacent categories are grouped yielding a smaller set of response categories, although that might be an attractive feature of a model, see also Jansen and Roskam (1986) where the extreme case of dichotomization in the polytomous Rasch model is considered in detail.

The cumulative or graded response model holds also for dichotomized responses (or other groupings of adjacent categories) since the splits themselves follow a binary model. This is different for the adjacent categories and the sequential model. They are not robust against collapsing of categories because the binary models that are contained are conditional, and collapsing of categories changes the conditioning. If, for example, categories r and $r+1$ are collapsed to form a new category, the conditions in the binary submodels after collapsing differ from the conditions in the original set of categories for all conditions that contain the new category. Since the graded response model is not conditional there is no condition that could change, and the model is robust against collapsing of categories.

4.6 Nominal Models and a Family of Models

The IRT models considered in the previous sections use the ordering of categories, which allows finding parsimonious parameterizations. In the literature also nominal models have been propagated that do not use the ordering but can be modified to do so. A specific family of models can be derived from Bock's nominal model (Bock, 1972).

Bock's Nominal Model

$$P(Y_{pi}=r)=\frac{\exp(\alpha_{ir}(\theta_p-\beta_{ir}))}{\sum_{s=0}^{k}\exp(\alpha_{is}(\theta_p-\beta_{is}))},\quad r=0,\dots,k. \quad (4.28)$$

It can also be given by

$$P(Y_{pi}=r)=\frac{\exp(\alpha_{ir}\theta_p-\tilde{\beta}_{ir})}{\sum_{s=0}^{k}\exp(\alpha_{is}\theta_p-\tilde{\beta}_{is})},\quad r=0,\dots,k,$$

where $\tilde{\beta}_{ir}=\alpha_{ir}\beta_{ir}$. In the original formulation, the parameterization $\alpha_{ir},\tilde{\beta}_{ir}$ was used but the parameterization α_{ir},β_{ir} yields the same model. To obtain identifiability of parameter constraints as, for example, $\alpha_{i0}=\tilde{\beta}_{i0}=0$ are needed, a more detailed discussion of identifiability conditions is found in Thissen and Cai (2016).

The model uses only the nominal scale level of the response. An indicator is that it remains the same if the response categories are permuted, which means it is stable under permutations. It becomes a model for ordered categories only by imposing further constraints.

In his investigation of the existence of sufficient statistics Andersen (1977) considered the simplified version

$$P(Y_{pi}=r)=\frac{\exp(\phi_r(\theta_p-\beta_{ir}))}{\sum_{s=0}^{k}\exp(\phi_s(\theta_p-\beta_{is}))}=\frac{\exp(\phi_r\theta_p-\tilde{\beta}_{ir})}{\sum_{s=0}^{k}\exp(\sum_{l=1}^{s}\phi_s\theta_p-\tilde{\beta}_{is})},$$

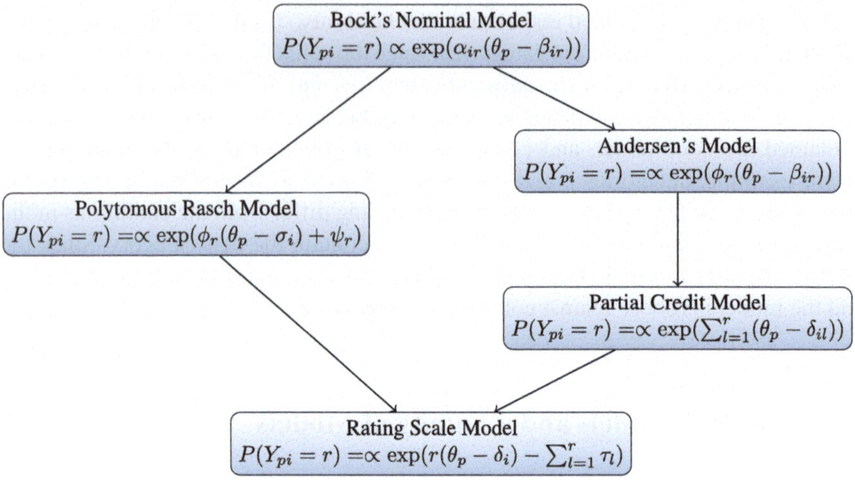

Fig. 4.22 Structure of polytomous latent trait models derived from Bock's model

where the scoring functions for the categories, α_{ir}, are replaced by $\alpha_{ir}=\phi_r$, which do not depend on the item. If, in addition, $\phi_r=r$ holds, the model simplifies to the partial credit model (4.17) (with parameters $\tilde{\beta}_{ir}=\sum_{l=1}^{r}\delta_{il}$).

Another simplified version of the nominal model is the unidimensional *Polytomous Rasch Model*

$$P(Y_{pi}=r)=\frac{\exp(\phi_r(\theta_p-\sigma_i)+\psi_r)}{\sum_{s=0}^{k}\exp(\phi_s(\theta_p-\sigma_i)+\psi_s)},$$

where $\alpha_{ir}=\phi_r$ and the decomposition $\beta_{ir}=\sigma_i+\gamma_r$ is assumed to hold. Then, one obtains the simplified version with $\psi_r=\phi_r\gamma_r$. Rasch (1961a) introduced the model as a unidimensional version of his most general multidimensional model. Andersen (1977) showed that the model has sufficient statistics only if for the difference of scoring functions $\phi_r-\phi_{r-1}=c$ holds for all r, where c is constant. For the special choice $\phi_r=r$, one obtains the rating scale model. The hierarchy of models visualized in Fig. 4.22 contains the two ordinal models that have been considered before in the section on adjacent categories models.

A common feature of the family of models given in Fig. 4.22 is that locally, given the response is in adjacent categories, one obtains 2PL-models. For Bock's general model, one obtains

$$P(Y_{pi}=r|Y_{pi}\in\{r-1,r\})=F((\alpha_{ir}-\alpha_{i,r-1})\theta_p-\delta_{ir}), \tag{4.29}$$

where $\delta_{ir}=\alpha_{ir}\beta_{ir}-\alpha_{i,r-1}\beta_{i,r-1}$ and $F(.)$ is the logistic function. Thus, locally a 2PL model holds with discrimination parameters $\alpha_{ir}-\alpha_{i,r-1}$. If the α-parameters are not restricted as in Bock's model, the probability of a response in category r given the response is in $\{r, r-1\}$ may increase with increasing θ_p but can also decrease with increasing θ_p if $\alpha_{ir} - \alpha_{i,r-1}$ is negative. The categories are not ordered with respect to θ_p and the model cannot be considered a sensible model for ordered responses.

Also the simplification by using scoring functions for the categories that do not depend on the item ($\alpha_{ir} = \phi_r$), which yields Anderson's model and the polytomous Rasch model, does not make the corresponding models ordinal models. They still do not use the ordering of categories. The essential postulate is that the scoring functions have to be ordered, that is, $\alpha_{i,r+1} \geq \alpha_{ir}$ has to hold for all r. It is of course fulfilled if $\phi_r = \alpha_{ir} = r$, which yields the partial credit model. As seen from Eq. (4.29), Bock's nominal model and therefore all models given in Fig. 4.22 can be seen as general adjacent categories model in which the discrimination parameters depend on the item *and* the category.

In general, if the objective is to measure a unidimensional trait like proficiency or attitude a nominal model is hardly sensible. The multinomial distribution (without any order restriction) calls for a multidimensional representation, k dimensions (parameters) are needed to represent a multinomial distribution with $k+1$ response categories. Models of this type have been considered by Takane and De Leeuw (1987) and Johnson and Bolt (2010). Johnson and Bolt (2010) refer to it as the factor-analytic multinomial logit model (FAMLM) given by

$$P(Y_{pi} = r) \propto \exp(\boldsymbol{\alpha}_{ir}^T \boldsymbol{\theta}_p + \beta_{ir}),$$

where $\boldsymbol{\alpha}_{ir}$ and $\boldsymbol{\theta}_p$ are k-dimensional vectors. The weight parameter $\boldsymbol{\alpha}_{ir}$ is item- and category-specific, and the latent trait is a multidimensional vector. The model is a latent trait version of the classical multinomial logit model, which is the dominating nominal model in multi-categorical regression analysis, see, for example, Tutz (2012), **?**. It can be derived by maximizing latent random utilities that follow a maximum extreme value distribution (Yellott, 1977; McFadden, 1973) and can be seen as a multidimensional generalization of Bock's nominal model. Johnson and Bolt (2010) used the model in particular for the incorporation of response styles, see also Bolt and Newton (2011). Alternative version of multidimensional models has been given by Thissen and Steinberg (2010), Thissen and Cai (2016).

Bock's model without order restrictions is indeed a nominal model. It holds, more generally, that the response probabilities given any two categories follow a 2PL model, more concrete, the model $P(Y_{pi} = m | Y_{pi} \in \{l, m\}) = F((\alpha_{im} - \alpha_{il})\theta_p - \delta_{im}))$, where $\delta_{im} = \alpha_{im}\beta_{im} - \alpha_{il}\beta_{il}$ holds. It depends on the difference $\alpha_{im} - \alpha_{il}$ if the probability of a response in category m increases or decreases if θ_p increases. This might be used to investigate in which way categories are ordered. Since the model is nominal with the implicit assumption that a unidimensional latent trait is present fitting the unconstrained model might yield information on the order of categories, provided the model holds, see, for example, Thissen and Cai (2016). When used in this way the model does not exploit the order of categories but investigates the order empirically. This is a different concept of dealing with ordinality, namely, using the model to examine the expected, or empirical, ordering of response categories instead of using the known order of categories to obtain sparse models.

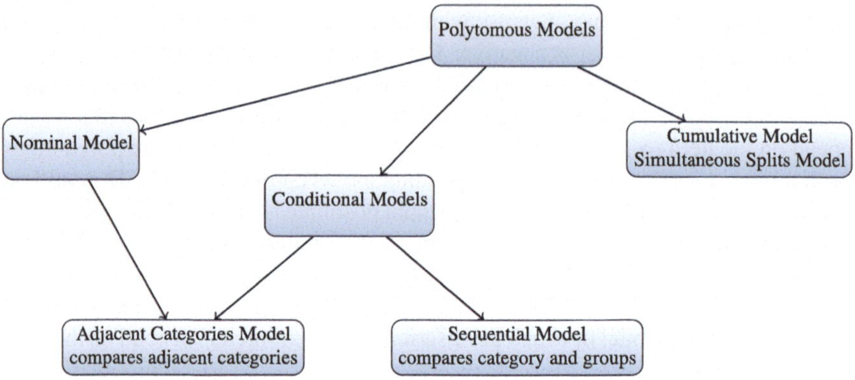

Fig. 4.23 Structure of polytomous latent trait models

The structure of polytomous models shown in Fig. 4.23 can be seen as a taxonomy of polytomous models. It is an extended version of Fig. 4.1, which shows ordinal models only. The nominal model, which is now included, is a polytomous model that is linked to the adjacent categories model but not to the two other ordinal model types, the sequential and the cumulative models.

4.7 Models for Nonmonotone Items

All the models considered in the previous sections are so-called dominance models in which a large latent trait means that high categories are more likely to be observed than low ones. This modeling approach is definitely appropriate in achievement tests with the latent trait being ability and the categories representing ordered categories of performance. It might be less appropriate for specific items in personality and attitude measurement.

As an example Lin et al. (2023) considered the item "I am sometimes organized and sometimes forgetful" in a test aiming at the measurement of conscientiousness. Respondents might disagree if they are well organized or very forgetful. Thus, a monotonic relationship between the probability of endorsement of the item and the underlying trait seems not warranted. Items of this type have been called double-barreled and of little value to differentiate in terms of their attitudes (Likert, 1932). However, also such items could be useful if modeled appropriately.

Models for items of this type are unfolding or ideal point models that assume a curvilinear relationship between the probability of endorsement of the item and the underlying personality trait. They assume an ideal point process in which a person endorses an attitude statement to the extent that it matches the person's opinion. It is assumed that there is a particular trait value, the ideal point, at which the probability of agreeing with the item peaks, and deviations from this point in either direction

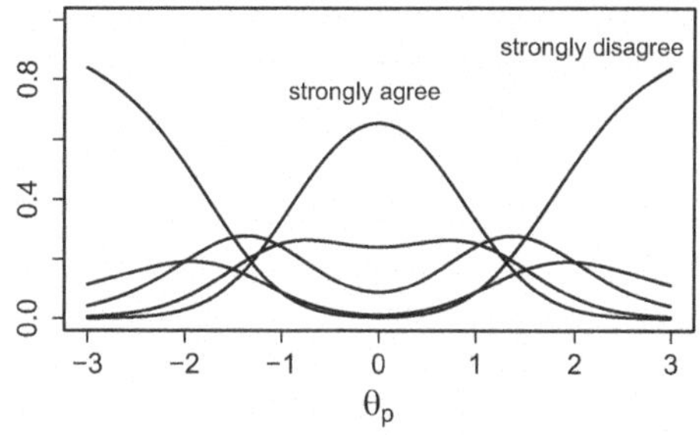

Fig. 4.24 Probabilities for the generalized grading unfolding model with five categories ranging from "strongly disagree" to "strongly agree"

lower the probability of endorsement (Roberts et al., 2000). The ideal point process and the unfolding technique for the formal representations of this process have been suggested by Coombs (1964) but has roots in the work of Thurstone (1929, 1928). Modern ideal point models have been proposed by Andrich (1988); Andrich and Luo (1993), Andrich (1996), Roberts et al. (2000).

A simple model proposed for binary responses by Andrich (1988) is

$$P(Y_{pi} = 1|\theta_p, \delta_i) = F(-(\theta_p - \delta_i)^2).$$

It is obvious that as the distance between the latent trait θ_p and the "ideal point" δ_i increases, the probability of an endorsed response decreases.

For polytomous models, the construction of a model is typically based on very specific assumptions on the latent process. To obtain the polytomous model proposed by Roberts et al. (2000), first, it is assumed that persons agree with the item to the extent that it is located close to their position on a unidimensional latent attitude continuum. Then, it is assumed that persons disagree strongly or weakly "from above" if the item is located far below the person's position on the attitude continuum. They disagree strongly or weakly "from below" if the item is located far above the person's position. Thus, for example, strong disagreement is represented by two latent categories, strong disagreement from below and strong disagreement from above. For the latent categories, considered subjective responses, a generalized partial credit model is assumed to hold. By assuming that the sum of the probabilities associated with the two corresponding subjective responses yields the observable response one obtains the generalized grading unfolding model (GGUM), which links latent traits to observations in the form

$$P(Y_{pi} = r|\theta_p, \delta_i, \alpha_i, \tau_{ir}) \propto \exp(\alpha_i(r(\theta_p - \delta_i) - \tau_{ir})) + \exp(\alpha_i(2k + 1 - r)(\theta_p - \delta_i) - \tau_{ir})),$$

and contains the item location parameters δ_i, τ_{ir} and discrimination parameters α_i. For illustration Fig. 4.24 shows the probabilities of a model with five categories ranging from "strongly disagree" to "strongly agree" for a model with $\delta_i = 0$. It is seen that the probabilities of a response in category r are the same for θ_p and $-\theta_p$, therefore the symmetric curves centered at $\delta_i = 0$. At the "ideal point" $\delta_i = 0$, the probability of strong agreement has its highest value, while the probability of strong disagreement is high if θ_p is far away from the ideal point. The response curves are quite different from response curves seen in dominance models, which are increasing or decreasing for the extreme categories. More concrete, they are decreasing for category 0 and increasing for category k as θ_p increases.

The usefulness of unfolding or ideal point models is linked to the type of items that are used. For vague or double-barreled items like "I am sometimes organized and sometimes forgetful", the model seems a good choice. If tests contain unambiguous items as tests often do it is less clear if they are modeled adequately. In a test that aims at measuring attitude toward abortion as considered by Roberts et al. (2000) the item "My feelings about abortion are very mixed" is a clear candidate for the unfolding model. It is less clear for unambiguous items like "Abortion is inhumane" or "Abortion should be legal under any circumstances". Since graded unfolding models assume that the probability of strong disagreement should be large if the person's attitude is far away from the ideal point, that is, above or below the ideal point it is assumed that the probability of strong disagreement should be large if a positive attitude toward abortion is very low or very high, which seems less plausible. For a discussion of the usefulness of unfolding models in the measurement of attitudes and personality, see Drasgow et al. (2010) and the more critical responses on the article, which favors unfolding models (Reise, 2010; Oswald and Schell, 2010).

Models as the generalized grading unfolding model are ordinal models in the sense that they exploit the order of response categories but follow a quite different paradigm than dominance models.Though unfolding models typically use classical ordinal models in the construction of the model they form a distinct class of models, which could be included in a taxonomy of models as a separate group. The structure shown in Fig. 4.23 refers to dominance only.

4.8 Estimation of Polytomous Models

One way to obtain estimates is marginal maximum likelihood estimation, which can be used for all models. The partial credit model allows to estimate parameters by conditional maximum likelihood since sufficient statistics of item parameters exist. A special estimation procedure is available for sequential models, for which parameters can be estimated by using estimation tools for binary models.

4.8.1 *Marginal Maximum Likelihood Estimation*

Marginal maximum likelihood (MML) estimates of item parameters use a fixed distribution of item parameters. It is useful to represent the responses $Y_{pi} \in \{0, 1, \dots, k\}$ by indicator variables defined by

$$Y_{pir} = \begin{cases} 1 & Y_{pi} = r \\ 0 & Y_{pi} \neq r. \end{cases}$$

The 0–1 indicator variables represent the responses in vector form:

$$Y_{pi} = r \quad \Leftrightarrow \quad \boldsymbol{Y}_{pi}^T = (Y_{pi0}, \ldots, Y_{pik}) = (0, \ldots, 0, 1, 0, \ldots, 0),$$

with the vector $\boldsymbol{Y}_{pi}$ containing just one entry "1". The indicator variables are strongly linked to the split variables, which have been defined by $Y_{pi}^{(r)} = 1$ if $Y_{pi} \geq r$, $Y_{pi}^{(r)} = 0$ if $Y_{pi} < r$. It is easily seen that $Y_{pi}^{(r)} = \sum_{j=r}^{k} Y_{pij}$ and $Y_{pir} = Y_{pi}^{(r)} - Y_{pi}^{(r+1)}$, where $Y_{pi}^{(k+1)} = 0$. Split variables and 0–1 indicator variables are just two alternative ways to represent observations in vector form.

With indicator variables the marginal likelihood has the form:

$$L_m = \prod_{p=1}^{P} \int \prod_{i=1}^{I} \prod_{r=0}^{k} P(Y_{pi} = r)^{y_{pir}} f(\theta_p) d\theta_p,$$

where $f(\theta_p)$ is the density of the distribution assumed for the person parameters. Typically, a normal distribution with mean zero and variance σ_θ^2, indicated by $f_{0,\sigma_\theta}(.)$, is assumed. The corresponding log-likelihood is

$$l_m = \log(L_m) = \sum_{p=1}^{P} \log(\int \prod_{i=1}^{I} \prod_{r=0}^{k} P(Y_{pi} = r)^{y_{pir}} f_{0,\sigma_\theta}(\theta_p) d\theta_p).$$

4.8.2 Conditional Estimation for Partial Credit Model

The probability of the responses for person p with total score $y_{p+} = \sum_i y_{pi} = r$ is given by

$$P(Y_{p1} = y_{p1}, \ldots, Y_{pI} = y_{pI}) = P(Y_{p1} = y_{p1}) \ldots P(Y_{pI} = y_{pI}) =$$

$$= \exp(\sum_{i=1}^{I} \sum_{l=1}^{y_{pi}} (\theta_p - \delta_{il}))/\tau(\theta_p) = \exp(r\theta_p - \sum_{i=1}^{I} \sum_{l=1}^{y_{pi}} \delta_{il})/\tau(\theta_p),$$

where $\tau(\theta_p) = (\sum_{s=0}^{k} \exp(\sum_{l=1}^{s} (\theta_p - \delta_{il})))^I$. For the distribution of the sum score $y_{p+} = r$, one obtains

$$P(\sum_{i=1}^{I} Y_{pi} = r) = \sum_{y_{p1}+\dots+y_{pI}=r} P(Y_{p1} = y_{p1}) \dots P(Y_{pI} = y_{pI})$$
$$= e^{r\theta_p} \sum_{y_{p1}+\dots+y_{pI}=r} e^{-\sum_{i=1}^{I}\sum_{l=1}^{y_{pi}} \delta_{il}} / \tau(\theta_p) = \qquad = e^{r\theta_p} \gamma_r(\boldsymbol{\delta})/\tau(\theta_p),$$

where $\gamma_r(\boldsymbol{\delta}) = \sum_{y_{p1}+\dots+y_{pI}=r} e^{-\sum_{i=1}^{I}\sum_{l=1}^{y_{pi}} \delta_{il}}$ is a function that does depend on item parameters only (and on r) but not on the person parameter. Thus, the conditional distribution of item responses given $y_{p+} = r$ is

$$P(Y_{p1} = y_{p1}, \dots, Y_{pI} = y_{pI} | \sum_{i=1}^{I} Y_{pi} = r) = \frac{\exp(-\sum_{i=1}^{I}\sum_{l=1}^{y_{pi}} \delta_{il})}{\gamma_r(\boldsymbol{\delta})},$$

which does not depend on the person parameter. The corresponding conditional likelihood has the form:

$$L_c = \prod_{p=1}^{P} P(Y_{p1} = y_{p1}, \dots, Y_{pI} = y_{pI} | \sum_{i=1}^{I} Y_{pi} = y_{p+}) = \prod_{p=1}^{P} \frac{\exp(-\sum_{i=1}^{I}\sum_{l=1}^{y_{pi}} \delta_{il})}{\gamma_{y_{p+}}(\boldsymbol{\delta})}. \quad (4.30)$$

4.8.3 Estimation of the Sequential Model Using Binary Variables

The probabilities of the sequential model are given by

$$P(Y_{pi} = r) = (1 - F(\alpha_{i,r+1}(\theta_p - \delta_{i,r+1}))) \prod_{j=1}^{r} F(\alpha_{ij}(\theta_p - \delta_{ij})), \quad r = 0, \dots, k,$$

where $\delta_{i,k+1} = \infty$. It can also be represented as the probability of a sequence of binary pseudo-items $(Y_{pi}^{(1)}, \dots, Y_{pi}^{(r+1)})^T = (1, \dots, 1, 0)$ with each of the pseudo-items following a binary model, more concise, $P(Y_{pi}^{(j)} = 1) = F(\alpha_{ij}(\theta_p - \delta_{ij})$. This is seen from

$$P(Y_{pi} = r) = P(Y_{pi}^{(r+1)} = 0) \prod_{j=1}^{r} P(Y_{pi}^{(j)} = 1).$$

The pseudo-items are nothing else than the split variables that have been considered before, but with the artificial split variable $Y_{pi}^{(k+1)}$ that is fixed by $Y_{pi}^{(k+1)} = 0$ and has item parameter $\delta_{i,k+1} = \infty$, which is not estimated. The number of pseudo-items that is used in the estimation depends on the observation itself. If category r

is observed the sequence contains $r + 1$ pseudo-items. It is helpful to distinguish between the cases $r < k$ and $r = k$, since in the latter case the artificial last pseudo-item can be dropped.

More concrete, for person p, item i with observation $r < k$ one uses the following sequence of pseudo-items:

	Item i			
Observations	$Y_{pi}^{(1)} = 1$	...	$Y_{pi}^{(r)} = 1$	$Y_{pi}^{(r+1)} = 0$
Parameters	δ_{i1}	...	δ_{ir}	$\delta_{i,r+1}$

For person p, item i with observation k one uses

	Item i		
Observations	$Y_{pi}^{(1)} = 1$	...	$Y_{pi}^{(k)} = 1$
Parameters	δ_{i1}	...	δ_{ik}

The corresponding marginal log-likelihood is given by

$$L_m = \prod_{p=1}^{P} \int \prod_{i=1}^{I} (1 - F(\alpha_{i,y_{pi}+1}(\theta_p - \delta_{i,y_{pi}+1})))^{Ind(y_{pi}<k)} \prod_{j=1}^{y_{pi}} F(\alpha_{ij}(\theta_p - \delta_{ij})) f(\theta_p) d\theta_p,$$

where $Ind(y_{pi} < k) = 1$ if $y_{pi} < k$ and $Ind(y_{pi} < k) = 0$ otherwise. The corresponding term is omitted if category k is observed. $f(\theta_p)$ is the density of the distribution assumed for the person parameters.

Together with missing values

	Item 1				Item I		
Observations	$Y_{p1}^{(1)}$	...	$Y_{p1}^{(k)}$	...	$Y_{pI}^{(1)}$	...	$Y_{pI}^{(k)}$
Parameters	δ_{11}	...	δ_{1k}	...	δ_{I1}	...	δ_{Ik}

4.9 Further Examples

Example 4.4 FIMS study

The FIMS dataset has been considered in previous examples to illustrate parameter estimation but goodness of fit has not been evaluated. Here, we briefly compare the classical ordinal models with regard to their goodness of fit measured by the log-likelihood. Table 4.9 shows the marginal log-likelihood for the six items that have been used in the previous examples. The models that have been fitted are the logistic graded response model, the partial credit model, and the logistic sequential model. In the upper part of the discrimination parameter was fixed ($\alpha_i = 1$). Then, all models have the same number of parameters. In addition, the generalized partial credit model (GPCM) has been fitted, in which the discrimination parameters vary freely.

Table 4.9 Model fits for PCM, logistic graded response model, and logistic sequential model (six items)

Model	Log-lik	Parameters
Logistic graded response model (α fixed)	−1285.830	31
Partial credit model (α fixed)	−1380.848	31
Logistic sequential model (α fixed)	−1493.023	31
GPCM	−1285.083	36

Table 4.10 Model fits for all items

Model	Log-lik	Parameters
Logistic sequential model (α fixed)	−2986.693	58
Partial credit model (α fixed)	−3110.556	58

It is seen that among the models with fixed discrimination parameters the logistic graded response model yields the best fit, followed by the PCM and the sequential model. The fit of the generalized PCM is as good as the fit of the graded response model but uses much more parameters. PCM was fitted by using function gpcm, graded response model by grm from package ltm (Rizopoulos, 2006; Johnson, 2007). The sequential model was fitted by own software.

Although the graded response models show the best fit when using only six items the model is not without problems since it contains a restriction that postulates the ordering of thresholds. If all 14 available items are included not all item parameters can be estimated. Estimates could only be found after collapsing categories in some items. Problems of this type do not occur for the sequential model and the PCM. Table 4.10 shows that the sequential model fits better than the graded response model if all items are included. The estimates of graded response model and the generalized PCM did not converge. □

Example 4.5 Fears Data
As an additional example, we consider data from the German Longitudinal Election Study (GLES), which have also been used in Example 2.4. The data originate from the pre-election survey for the German federal election in 2017 and are concerned with political fears. The participants were asked: "How afraid are you due to the ..." (1) refugee crisis? (2) global climate change? (3) international terrorism? (4) globalization? (5) use of nuclear energy? The answers were measured on Likert scales from 1 (not afraid at all) to 7 (very afraid) and the model is fitted under the assumption that fear is the dominating latent trait, which is considered as unidimensional. We use 200 persons sampled randomly from the available set of observations. Since the sequential model as a process model of consecutive decisions is less appropriate for attitude data we restrict attention to the graded response model and the partial credit model. Table 4.11 shows the corresponding log-likelihoods. It is seen that the graded

Table 4.11 Log-likelihood values for fears data

Model	Log-lik	Parameter
Logistic graded response model (α fixed)	−1707.573	31
Logistic generalized graded response model	−1700.555	36
Partial credit model (α fixed)	−1742.285	31

response model fits better than the partial credit model. An even better fit is seen for the graded response model with discrimination parameters, the corresponding generalized PCM did not converge. □

4.10 Fitting of Models

Marginal estimates for the Graded Response, and the Generalized Partial Credit Models can be obtained by using R package *ltm* (Rizopoulos, 2006).

The R package *TAM* (Robitzsch et al., 2024) can be used to obtain marginal maximum likelihood estimation and joint maximum likelihood estimation for the generalized partial credit model, the nominal item, and mixture distribution IRT models.

The package *eRm* (Mair and Hatzinger, 2007b, a) provides conditional fitting of the rating scale model, linear rating scale models, partial credit models, and linear partial credit models (LPCM). For a more general description of the available models that can be fitted by these programs, see Chap. 12.

The package *GGUM* Tendeiro and Castro-Alvarez (2019) allows to fit the generalized graded unfolding model proposed by Roberts et al. (2000).

Chapter 5
Extended Ordinal Models: Accounting for Response Styles

Response styles are an important problem in psychological measurement since, if present, they can affect the accuracy of measurement. In Likert-type scales, which represent the level of agreement in the form *strongly disagree, moderately disagree,..., moderately agree, strongly agree* interesting response styles are the extreme response style, which means persons have a tendency to extreme categories, and its counterpart, the tendency to favor middle categories. The problem with response styles is that they can affect the validity of scale scores because estimates of the substantive trait may be biased if the response style is ignored. Models that explicitly account for response styles are able to reduce the bias. They account for additional heterogeneity in the population, which in some applications is itself of interest, in particular if it is linked to explanatory variables. In item response data respondents rate their level of agreement on a series of items and response styles can be considered a *consistent pattern of responses that is independent of the content of a response* (Johnson, 2003). Various response styles have been identified, for an overview see, for example, Messick (1991), Baumgartner and Steenkamp (2001).

In the following some concepts how to include response styles into basic models like the graded response model and the partial credit model by adding response style parameters are given. The inclusion of response style parameters yields multi-trait models, which contain more than one person parameters.

5.1 Tendency to Middle or Extreme Categories in Partial Credit Models

Let the categories $0, \dots, k$ represent graded agree-disagree attitudes with a natural symmetry like *strongly disagree, moderately disagree,..., moderately agree, strongly agree*. In the partial credit model the predictor, when choosing between categories $r-1$ and r, has the form $\eta_{pir} = \theta_p - \delta_{ir}$. The parameter δ_{ir} determines the choice

G. Tutz, *A Short Guide to Item Response Theory Models*, Statistics for Social and Behavioral Sciences, https://doi.org/10.1007/978-3-031-87271-6_5

between categories $r-1$ and r. Response styles that account for a tendency to extreme categories or a tendency to middle categories are modeled by modifying the thresholds δ_{ir}. In the predictor, an additional person parameter γ_p is included that shifts the thresholds of the agreement categories and the thresholds of the disagreement categories into opposite directions.

Let $m = k/2$ denote the middle of the response categories. If the number of response categories is odd (k even) m is the neutral middle category between disagreement categories $0, \ldots, m-1$ and agreement categories $m+1, \ldots, k$. If, for example, $k = 4$ the levels of disagreement are 0, 1, the levels of agreement are 3, 4 with $m = 2$ representing the middle category. If the number of categories is even (k odd) the responses are divided into disagreement categories $0, \ldots, [m]$ and disagreement categories $[m]+1, \ldots, k$, where $[.]$ denotes the floor function. If, for example, $k = 3$ the disagreement categories are 0, 1 and the agreement categories are 2, 3. Then, $m = 1.5$ denotes the middle of $0, \ldots, k$ but is not itself a response category.

In the common partial credit model the predictor, when choosing between categories $r-1$ and r, has the form $\eta_{pir} = \theta_p - \delta_{ir}$, and the parameter δ_{ir} determines the choice between categories $r-1$ and r. Response styles that account for a tendency to extreme categories or a tendency to middle categories can be modeled by modifying the thresholds δ_{ir}. In the predictor an additional person parameter γ_p is included that shifts the thresholds of the agreement categories and the thresholds of the disagreement categories into opposite directions. The corresponding *partial credit model with response style* (PCMRS) as proposed by Tutz et al. (2018) is given by

$$\log\left(\frac{P(Y_{pi}=r)}{P(Y_{pi}=r-1)}\right) = \theta_p + (m-r+0.5)\gamma_p - \delta_{ir} = \theta_p - \tilde{\delta}_{ir}, \quad r = 1, \ldots, k \tag{5.1}$$

with the modified thresholds defined by $\tilde{\delta}_{ir} = \delta_{ir} - (m-r+0.5)\gamma_p$. For illustration let us consider the case of five response categories ($k = 4, m = 2$). Then the response categories are partitioned into the disagreement categories $\{0, 1\}$, the neutral category $\{2\}$ and the agreement categories $\{3, 4\}$, and the modified thresholds are given by $\tilde{\delta}_{i1} = \delta_{i1} - 1.5\gamma_p$, $\tilde{\delta}_{i2} = \delta_{i2} - 0.5\gamma_p$, $\tilde{\delta}_{i3} = \delta_{i3} + 0.5\gamma_p$, $\tilde{\delta}_{i4} = \delta_{i4} + 1.5\gamma_p$. One obtains the following structure of categories and thresholds

If $k = 3$ one has $m = 1.5$ and the modified thresholds are given by $\tilde{\delta}_{i1} = \delta_{i1} - \gamma_p$, $\tilde{\delta}_{i2} = \delta_{i2}$, $\tilde{\delta}_{i3} = \delta_{i3} + \gamma_p$ yielding the following structure of categories and thresholds

The parameter γ_p generates a shifting of thresholds. If γ_p is positive one has for the disagreement categories a shifting of the thresholds δ_{ir} to the left, for the agreement categories a shifting to the right. The corresponding new thresholds are given by $\tilde{\delta}_{ir} = \delta_{ir} - (m-r+0.5)\gamma_p$. The scaling factor $(m-r+0.5)$ on the parameter γ_p determines the direction of the shifting and is chosen such that the difference between the modified thresholds of adjacent categories changes by a constant,

$$\tilde{\delta}_{ir} - \tilde{\delta}_{i,r-1} = \delta_{ir} - \delta_{i,r-1} + \gamma_p. \tag{5.2}$$

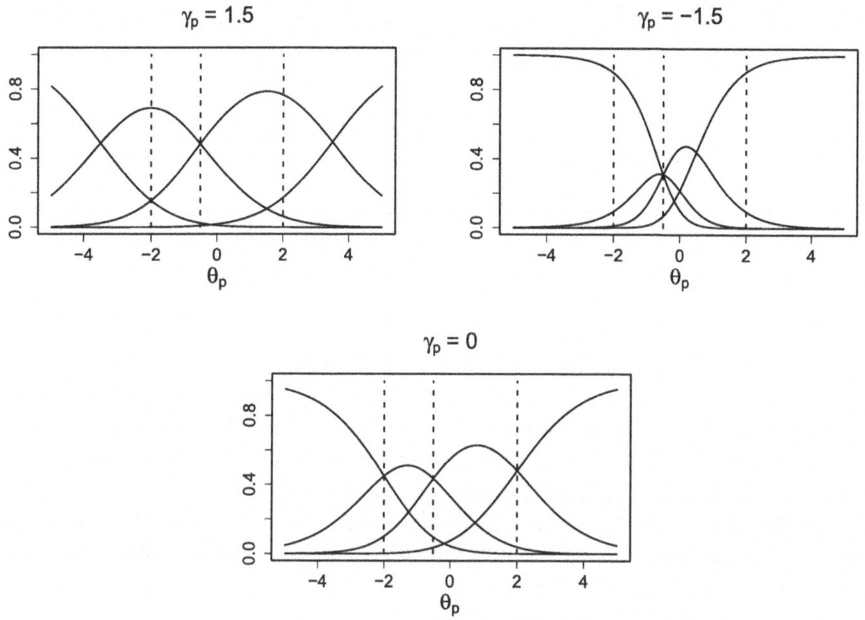

Fig. 5.1 Probabilities $P(Y_{pi} = r)$ against θ_p for positive, negative γ_p and $\gamma_p = 0$; number of categories is four

All the differences become larger for positive γ_p and smaller for negative γ_p. For positive γ_p the probability mass is shifted to the middle categories, for negative γ_p it is shifted to the extreme categories. It is crucial that the modification of the thresholds is done in a symmetric way with a centering in the middle of the response scale.

Positive γ_p means the person has a tendency to middle categories and negative γ_p indicates that the person has a tendency to extreme categories.

The role of the response style parameter becomes especially obvious when considering extreme values. If the number of response categories is odd the extreme case $\gamma_p \to \infty$ yields $P(Y_{pi} = m) \to 1$. Then the whole probability mass is in the middle category. For $\gamma_p \to -\infty$ the whole probability mass is in the extreme categories 0 and k. Thus, positive γ_p indicates that the person has a tendency to middle categories and negative γ_p indicates that the person has a tendency to extreme categories. Figure 5.1 visualizes the impact of the response style parameter for $k = 3$. It is seen that for positive γ_p the probability for middle categories increases, while it increases for extreme categories if γ_p is negative.

The PCMRS is an extended partial credit model that contains not only one but two person parameters, one for the latent trait that is of primary interest and one for the response style. The main reason why it is useful to account for response styles is that the accuracy of parameter estimates can suffer strongly if response styles are ignored, see Tutz et al. (2018).

Also for the PCMRS model the specific objectivity or separability property holds. For the comparison of items i and j one obtains again

$$\log\left(\frac{P(Y_{pi}=r)}{P(Y_{pi}=r-1)}\right)-\log\left(\frac{P(Y_{pj}=r)}{P(Y_{pj}=r-1)}\right)=-(\delta_{ir}-\delta_{jr}),$$

which does not depend on the person parameter θ_p or the response style parameter γ_p. Therefore, the comparison of item difficulties does not depend on the persons. For the comparison between person p and person $\tilde{p}$ one obtains

$$\log\left(\frac{P(Y_{pi}=r)}{P(Y_{pi}=r-1)}\right)-\log\left(\frac{P(Y_{\tilde{p}i}=r)}{P(Y_{\tilde{p}i}=r-1)}\right)=\theta_p-\theta_{\tilde{p}}+(m-r+0.5)(\gamma_p-\gamma_{\tilde{p}}),$$

which does not depend on the item parameters. Therefore, the comparison of persons is just a function of the person parameters, which now includes the ability parameters $\theta_p, \theta_{\tilde{p}}$ and the response style parameters $\gamma_p, \gamma_{\tilde{p}}$.

The PCMRS can be seen as a multidimensional IRT model with clearly defined dimensions. General versions of multidimensional partial credit models have been considered by Kelderman (1996), Wetzel and Carstensen (2017), Plieninger (2016). Models of this form can be represented by

$$P(Y_{pi}=r)=\frac{\exp(\sum_{l=1}^{r}(\sum_{d=1}^{D}\omega_{ird}\theta_{pd}-\delta_{il}))}{\sum_{s=0}^{k}\exp(\sum_{l=1}^{s}(\sum_{d=1}^{D}\omega_{ird}\theta_{pd}-\delta_{il}))},\quad r=1,\dots,k,$$

yielding

$$\log\left(\frac{P(Y_{pi}=r)}{P(Y_{pi}=r-1)}\right)=\eta_{pir}=\sum_{d=1}^{D}\omega_{ird}\theta_{pd}-\delta_{ir},\quad r=1,\dots,k,$$

where the person parameter $\boldsymbol{\theta}_p^T=(\theta_{p1},\dots,\theta_{pD})$ is now a vector with D dimensions, and ω_{ird} are scoring weights. If $D=1$ and for all items $\omega_{ird}=1$, one obtains the simple partial credit model. The PCMRS is obtained if $d=2$ with person parameters $(\theta_{p1},\theta_{p1})=(\theta_p,\gamma_p)$ and scoring weights $(\omega_{ir1},\omega_{ir1})=(1,m-r+0.5)$. An advantage of low-dimensional models like the PCMRS is that dimensions are clearly defined as substantive trait and response style, and there is no need to search for an interpretation of multi-dimensional latent traits that are just assumed to be present without further specification.

Marginal estimation of item parameters, which is used in the examples, uses that person parameters are normally distributed with expectation zero and covariance matrix $\boldsymbol{\Sigma}$. In contrast to simple ordinal models one has to assume a two-dimensional distribution of person parameters since one has a vector of person parameters, (θ_p,γ_p), for details, see Sect. 5.3.

Example 5.1 Science
The data set comes from the Consumer Protection and Perceptions of Science and Technology section of the 1992 Euro-Barometer Survey (Reif and Melich (1995)) based on a sample from Great Britain and is included in the R package *ltm*. All the items were measured on a four-group scale with response categories "0: strongly disagree", "1 :disagree to some extent", "2: agree to some extent" and "3: strongly agree". The items as described in the package *ltm* are:
(1) Comfort: Science and technology are making our lives healthier, easier, and more comfortable.
(2) Environment: Scientific and technological research cannot play an important role in protecting the environment and repairing it.
(3) Work: The application of science and new technology will make work more interesting.
(4) Future: Thanks to science and technology, there will be more opportunities for the future generations.
(5) Technology: New technology does not depend on basic scientific research.
(6) Industry: Scientific and technological research do not play an important role in industrial development.
(7) Benefit: The benefits of science are greater than any harmful effect it may have.
In most items, the agreement and disagreement categories are ordered such that large numbers indicate a positive attitude toward science. The exceptions are the items 2, 5, and 6. Therefore, the ordering for these items has been reversed before the fitting of models.

Fitting of the extended partial credit model and printing of the parameters can be obtained by using the package *PCMRS* (Schauberger, 2022):

```
library("PCMRS")
ScienceReorderedframe<-readRDS("ScienceReorderedframe")
respfit <- PCMRS(ScienceReorderedframe)

round(respfit$delta, digits=3)
respfit$Sigma
round(respfit$delta.PCM, digits=3)
respfit$sigma.PCM
```

The output shows the estimates of item parameters and the mixing distribution of person parameters for the extended PCM

```
               Catgr.1 Catgr.2 Catgr.3
Comfort         -2.597  -2.286   1.591
Environment     -0.337   0.542   1.798
Work            -1.686  -0.813   1.980
Future          -2.262  -1.201   1.248
Technology      -0.467   0.624   2.392
Industry        -0.185   1.438   2.341
Benefit         -2.174  -0.750   1.463

            theta      gamma
theta  0.2893575 -0.1100445
gamma -0.1100445  1.8266186
```

It is seen from the covariance matrix of the mixing distribution that the estimated variance of the response style parameter is even larger than the variance of the θ- parameter. Therefore, it seems not advisable to ignore the response style. The estimated covariance between θ and γ parameters is negative but very small. The corresponding estimates for the simple PCM are

```
               Catgr.1 Catgr.2 Catgr.3
Comfort         -2.258  -2.272   1.189
Environment     -0.248   0.576   1.460
Work            -1.371  -0.796   1.503
Future          -1.987  -1.186   0.903
Technology      -0.335   0.657   1.970
Industry        -0.127   1.477   1.969
Benefit         -1.870  -0.737   1.060

[1] 0.269361
```

It is seen that parameter estimates typically are closer to zero if response styles are neglected. For illustration, in Fig. 5.2 exemplarily parameter estimates for two items are shown (drawn lines refer to PCMRS, dashed to PCM).

Alternatively the model can also be fitted by using the more flexible package *MultOrdRS* (Schauberger, 2024b):

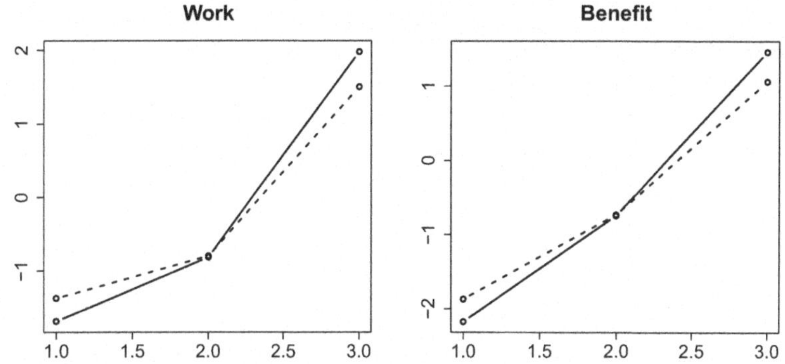

Fig. 5.2 Parameter estimates for PCMRS (drawn) and PCM (dashed)

```
library("MultOrdRS")
f.tense0 <-
  as.formula(paste("cbind(",paste(colnames(ScienceReorderedframe),
  collapse=","),") ~ 1"))
pcmrespst <- multordRS(f.tense0, data = ScienceReorderedframe)
round(-pcmrespst$beta.thresh, digits=3)
```

To avoid confusion, it should be noted that the function multordRS uses a different parameterization and yields parameters $-\delta_{ir}$ instead of δ_{ir}. □

The tendency to choose extreme or middle categories is not restricted to symmetric responses as found in Likert scales. It can also be present if responses are on a non-symmetric scale as the following example illustrates.

Example 5.2 Tenseness
We use data from the standardization sample of the Freiburg Complaint Checklist (FCC) (Fahrenberg, 2010; ZPID, 2013). The FCC is a questionnaire that is used to assess physical complaints of adults. The revised version of the FCC contains 71 items that can measure complaints on 9 different scales, as for example the scales *general condition*, *tenseness*, or *emotional reactivity*. The primary data of the standardization sample contains data on 2070 participants. Each of the 71 items is measured on a 5-point response scale that refers to the frequency of the complaint: "never", "about 2 times a year", "about 2 times a month", "approximately 3 times a week", or "almost every day".

The tenseness data contain 8 items for 2042 participants. In the 5-point scale low numbers correspond to low frequencies or low intensities of the following complaints: Clammy hands: Do you have clammy hands? Sweat attacks: Do you have sudden attacks of sweating? Clumsiness: Do you notice that you behave clumsy? Wavering hands: Are your hands wavering frequently, e.g., when lighting a cigarette or when holding a cup? Restless hands: Do you notice that your hands are restless? Restless feet: Do you notice that your feet are restless? Twitching eyes: Do you notice involuntary twitching of your eyes? Twitching mouth: Do you notice involuntary twitching of your mouth? The data are included in the package *PCMRS*, which can be used to estimate the model. In the following only the first four items are used.

```
library("PCMRS")
data(tenseness)
respfit <- PCMRS(tenseness[,1:4])
```

The parameter estimates for the PCMRS are

```
                Catgr.1 Catgr.2 Catgr.3 Catgr.4
Clammy hands     -1.016  -0.677  -0.218  -2.137
Sweat attacks    -0.823  -1.051  -0.363  -1.694
Clumsiness       -1.250  -1.360  -0.691  -1.997
Wavering hands   -0.921  -0.764  -0.595  -2.305
```

with estimated covariance matrix

```
       theta   gamma
theta 0.300    0.168
gamma 0.168    0.892
```

The parameter estimates for the PCM are again closer to zero than in the extended model:

```
                 Catgr.1 Catgr.2 Catgr.3 Catgr.4
Clammy hands      -0.807  -1.183  -0.090  -1.839
Sweat attacks     -0.557  -1.516  -0.245  -1.452
Clumsiness        -1.006  -1.881  -0.586  -1.715
Wavering hands    -0.777  -1.315  -0.466  -1.964
```

with estimated variance 0.372. □

5.2 Uncertainty in Partial Credit Models

An alternative response behavior is characterized by varying degrees of uncertainty. It means that respondents may respond in a deliberate way, knowing exactly which category they prefer, or suffer from a high degree of uncertainty, responding nonpurposefully. It is related to the *noncontingent response style* (NCR), which is found if persons have a tendency to respond to items carelessly, randomly, or nonpurposefully (Van Vaerenbergh and Thomas, 2013; Baumgartner and Steenkamp, 2001). Uncertainty can be modeled explicitly by introducing subject-specific parameters that are consistent throughout items and might be determined by external explanatory variables. The model explicitly aims at modeling the heterogeneity in the population. As demonstrated by Tutz and Schauberger (2020) ignoring heterogeneity may yield strongly misleading parameter estimates.

An extended version of the partial credit accounting for uncertainty is the *uncertainty partial credit model* (UPCM).

$$P(Y_{pi} = r) = \frac{\exp(\sum_{l=1}^{r} e^{\alpha_p}(\theta_p - \delta_{il}))}{\sum_{s=0}^{k} \exp(\sum_{l=1}^{s} e^{\alpha_p}(\theta_p - \delta_{il}))}, \quad r = 1, \ldots, k. \tag{5.3}$$

The usual predictor in the PCM, $\eta_{pir} = \theta_p - \delta_{ir}$, which distinguishes between category $r - 1$ and r, is replaced by the more general predictor

$$\eta_{pir} = e^{\alpha_p}(\theta_p - \delta_{ir}), \quad r = 1, \ldots, k,$$

which contains the additional subject-specific parameter α_p. The parameter α_p can be seen as modeling the subject-specific decisiveness or discriminatory power. For large α_p the person has distinct preferences, for small α_p the person tends to choose one of the response categories at random. Since it is not possible to determine if indecisiveness or carelessness is the reason the subject-specific effect e^{α_p} is referred to as *uncertainty effect*. Although the used terminology primarily refers to attitude measurement or personality questionnaires uncertainty may also come into play in achievement tests. The uncertainty may refer to a nonpurposeful response representing a person's ability to work concentrated or distractedly.

The impact of the parameter α_p is visualized in Fig. 5.3. It shows the response probabilities of the extended PCM with four categories for different values of α_p. It is seen that for small α_p one comes closer to a uniform distribution across categories, whatever the parameter θ_p is, for large α_p the preference for categories becomes very distinct depending on the value of θ_p. For $\alpha_p = 0$ one obtains the response probabilities given in the lower picture in Fig. 5.1. It represents the response probabilities for the traditional PCM without subject-specific heterogeneity in discrimination.

Parameters α_p can also be included if the response is binary, which corresponds to the special case $k = 1$. Following Tutz and Schauberger (2020) one has:

If $\alpha_p = 0$ for all p one obtains the binary Rasch model.

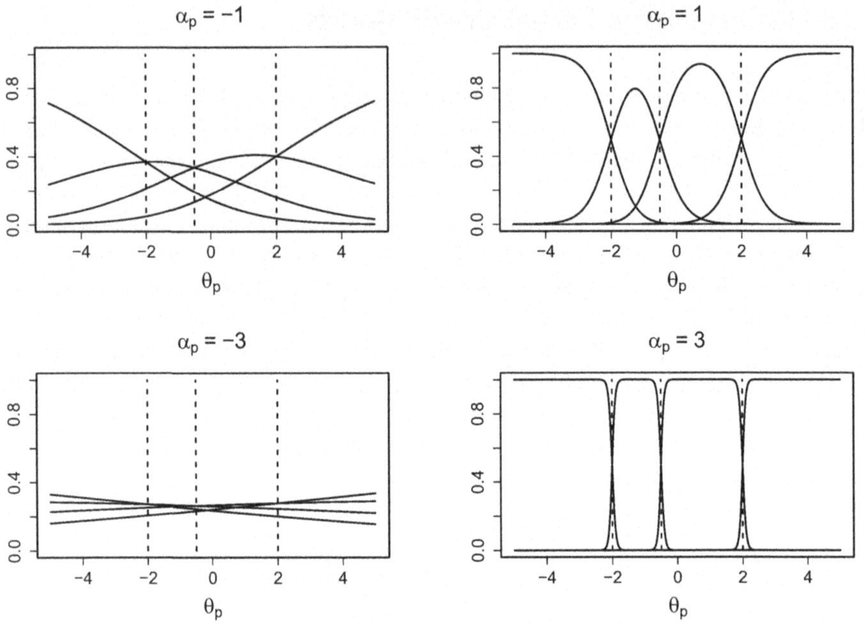

Fig. 5.3 Response probabilities in an extended PCM for several values of α_p (ordered thresholds)

If $\alpha_p > 0$ the person p is a strong discriminator and has a distinct preference for one of the two categories with the preference depending on θ_p. In particular, for $\alpha_p \to \infty$ one obtains $P(Y_{pi} = 1) = 1$ if $\theta_p > \delta_{i1}$, and $P(Y_{pi} = 0) = 1$ if $\theta_p < \delta_{i1}$.

If $\alpha_p < 0$ the person p is a weak discriminator. For $\alpha_p \to -\infty$ one obtains $P(Y_{pi} = 1) = 0.5$ for all abilities/attitudes θ_p. The person shows a noncontingent response style, which means the person has a tendency to respond to items randomly, or nonpurposefully.

In the general PCM it should be distinguished between two cases, ordered thresholds and un-ordered thresholds. If thresholds are *ordered*, that is $(\delta_{ir} \leq \delta_{i,r+1})$, one has:

If $\alpha_p = 0$ for all p one obtains the traditional PCM.

For $\alpha_p \to \infty$ one obtains for a person with $\theta_p \in (\delta_{ir}, \delta_{i,r+1})$ the probability $P(Y_{pi} = r) = 1$, the knows exactly which category he or she prefers. The property holds for all k if one defines $\delta_{i0} = -\infty$, $\delta_{i,k+1} = \infty$.

For $\alpha_p \to -\infty$ one obtains $P(Y_{pi} = r) = 1/(k + 1)$ for all abilities/attitudes θ_p. The person's response has a discrete uniform distribution over the response categories, which means random responding.

In the case of *reverse thresholds*, one finds a more distinct avoidance of the middle category than in the traditional PCM. For example, if $k = 2$ and *reverse thresholds* $\delta_{i2} < \delta_{i1}$ one sees the following:

For $\alpha_p \to \infty$ one obtains

For all persons $P(Y_{pi} = 1) = 0$, that is, the middle category is never chosen.

For persons with $\theta_p < (\delta_{i1} + \delta_{i2})/2$ one has $P(Y_{pi} = 0) = 1$.

For persons with $\theta_p > (\delta_{i1} + \delta_{i2})/2$ one has $P(Y_{pi} = 2) = 1$.

For $\alpha_p \to -\infty$ one obtains the same behavior as in the case of ordered thresholds.

Also the generalized partial credit model considered by Muraki (1992), Muraki (1997) can be extended to the *uncertainty generalized partial credit model* (UGPCM)

$$\log\left(\frac{P(Y_{pi} = r)}{P(Y_{pi} = r - 1)}\right) = e^{\alpha_p} a_i(\theta_p - \delta_{ir}), \quad r = 1, \ldots, k,$$

which includes an item-specific slope parameter a_i. In the generalized partial credit model the items have differing discriminatory power. The additional subject-specific uncertainty parameter means that discriminatory power varies also across persons.

Example 5.3 Tenseness

We use again the tenseness data from Example 5.2 and compare the uncertainty model and the partial credit model fits for the first four items.

```
library("UPCM")
data(tenseness)

#### uncertainty model
tensenessobs<-na.omit(tenseness)
Y <- data.matrix(tensenessobs[,1:4])

upcm <- UPCM(Y = Y, X = NULL, cores = 2, GPCM = FALSE)
round(upcm$delta, digits=3)
upcm$loglik

### PCM
pcmltm<-gpcm(tensenessobs[,1:4], constraint = "rasch")
pcmltm
```

Parameter estimates for the uncertainty model (log-likelihood −7196.579):

	Catgr.1	Catgr.2	Catgr.3	Catgr.4
Clammy_hands	1.841	0.095	1.137	0.552
Sweat_attacks	1.442	0.280	1.503	0.091
Clumsiness	1.706	0.696	2.921	0.318
Wavering_hands	2.131	0.374	1.509	0.581

with estimated covariance matrix

	theta	alpha
theta	0.194	0.338
alpha	0.338	1.321

Parameter estimates for the PCM (log-likelihood −7342.104):

	Catgr.1	Catgr.2	Catgr.3	Catgr.4	Dscrmn
Clammy_hands	1.866	0.331	1.576	1.239	1
Sweat_attacks	1.393	0.496	1.902	0.862	1
Clumsiness	1.751	0.837	2.316	1.406	1
Wavering_hands	1.986	0.743	1.706	1.189	1

5.3 Estimation

Estimation can be obtained by marginal maximum likelihood methods. Again the responses $Y_{pi} \in \{0, 1, \dots, k\}$ are represented by indicator variables defined by

$$Y_{pir} = \begin{cases} 1 & Y_{pi} = r \\ 0 & Y_{pi} \neq r. \end{cases}$$

Then, the marginal likelihood, which assumes that person parameters are normally distributed, has the form

$$L_m = \prod_{p=1}^{P} \int \prod_{i=1}^{I} \prod_{r=0}^{k} P(Y_{pi} = r)^{y_{pir}} f_{0,\boldsymbol{\Sigma}}(\theta_p, \gamma_p) d\theta_p d\gamma_p,$$

where $f_{0,\boldsymbol{\Sigma}}(.)$ is the density of the normal distribution with mean zero and covariance matrix $\boldsymbol{\Sigma}$. The corresponding log-likelihood is

$$l_m = \log(L_m) = \sum_{p=1}^{P} \log(\int \prod_{i=1}^{I} \prod_{r=0}^{k} P(Y_{pi} = r)^{y_{pir}} f_{.\mathbf{0},\boldsymbol{\Sigma}}(\theta_p, \gamma_p) dd\gamma_p.$$

The program packages *PCM* and *PCMRS* use Gauss-Hermite quadrature to obtain estimates.

5.4 Further Approaches to the Modeling of Response Styles

Other approaches to model response styles are based on Bock's nominal response model Bock (1972). Bolt and Johnson (2009), Johnson and Bolt (2010), Falk and Cai (2016) use extensions of the model to investigate the presence of a response style dimension. Bolt and Newton (2011) extend the approach of Bolt and Johnson (2009) to the analysis of rating data from multiple scales. Johnson (2003) considered a cumulative type model for extreme response styles using Bayesian methods.

An alternative strategy for measuring response style uses *finite mixtures*, which were introduced by Rost (1990). It is assumed that the observed response is a mixture of a finite number of latent responses, that means, the whole population can be divided into disjunctive latent classes. After classes have been identified it is investigated if item characteristics differ between classes, potentially revealing differing response styles, see, for example, Eid and Rauber (2000), Gollwitzer et al. (2005), Maij-de Meij et al. (2008), Moors (2010),Van Rosmalen et al. (2010). As Bolt and Johnson (2009) pointed out, in these models response style is viewed as a discrete qualitative difference in which each respondent is a member of one class. This is different from response styles as a continuous trait. An instructive overview on mixture-distribution and HYBRID Rasch models was given by von Davier and Yamamoto (2007).

A quite different strategy uses tree methodology to investigate response styles. Trees, in general, assume a nested structure where first a decision about the direction of the response and then about the strength is obtained, see Chap. 9 and the references therein.

5.5 Fitting of Models

The R package UPCM can be used to fit the partial credit Model with response styles. The model allows for an additional person parameter that characterizes the response style of the person.

The R package MultOrdRS provides methods to account for personal response styles and to explain the effects of covariates on the response style. The cumulative model (graded response model) as well as the adjacent categories model can be fitted.

Chapter 6
The Thresholds Model: A Common Framework for Discrete and Continuous Responses

In this chapter a general model is considered, which contains various latent trait models as special cases, in particular the binary models considered in this book and some polychotomous models. One strength of the overarching model is that it provides a genuine latent trait model for continuous responses, which may be restricted, for example, to be positive or in an interval. Continuous responses, as the visual analogue scale, have been in wide use for the measurement of quantities such as pain, quality of life and anxiety for quite some time and have gained in popularity owing to its ability to precisely and easily assess how people feel (Heller et al., 2016). The visual analogue scale consists of a line with anchor descriptors at both ends of the line and is an example of continuous data restricted to an interval, which means normal distributions are not appropriate.

The thresholds model for continuous responses is also as an alternative to classical test theory. The basic concepts of classical test theory (CTT) and the links between the model and CTT will be considered separately.

The thresholds model does not only contain other models as special cases it also allows to combine items with different formats within one test. Some items can have continuous responses, others binary or polytomous responses. The common model automatically accounts for the distributional differences. Thus, the model links person abilities to the performance on items that can differ in their response distribution.

After introducing the model we will focus on models for continuous responses but also consider Likert-type items. Models for discrete responses with infinite support will be treated in Chap. 8.

G. Tutz, *A Short Guide to Item Response Theory Models*, Statistics for Social and Behavioral Sciences, https://doi.org/10.1007/978-3-031-87271-6_6

6.1 The Thresholds Model

Let again Y_{pi} denote the response of person p on item i. However, the range of the values the response can take is not fixed in the way it was in the previous sections. More generally, it is assumed that Y_{pi} has support S. Interesting cases are $S = \{0, 1\}$ if responses are binary, $S = \{0, 1, \ldots, k\}$ for polychotomous responses, $S = \{0, 1, \ldots, \}$ for count data, $S = [0, \infty)$ for continuous, non-negative responses, for example response times, $S = (-\infty, \infty)$ for continuous, unrestricted responses.

Let, as before, $F(.)$ denote a strictly monotonically increasing distribution function. Then the general model has the following form.

Thresholds Model

$$P(Y_{pi} > y|\theta_p, \alpha_i, \delta_i(.)) = F(\alpha_i(\theta_p - \delta_i(y))), \quad y \in S. \quad (6.1)$$

As the models in previous sections it contains a person parameter θ_p and a discrimination parameter α_i. The main difference to previous models is that it contains an item-specific function $\delta_i(.)$, called *item difficulty function*, which is defined on the support S, and which is assumed to be *non-decreasing*. It is easy to see that for binary responses ($S = \{0, 1\}$) one obtains with $\delta_{i0} := \delta_i(0)$ the homogeneous binary model $P(Y_{pi} = 1|\theta_p, \alpha_i, \delta_i) = F(\alpha_i(\theta_p - \delta_{i0}))$. Since $P(Y_{pi} > 1|\theta_p, \alpha_i, \delta_i(.)) = 0$ the item difficulty function at 1 is given by $\delta_i(1) = \infty$.

The model is in particular determined by

- the response function $F(.)$, which is a strictly monotonically increasing distribution function and
- the difficulty function $\delta_i(.)$, which is a monotonically non-decreasing function.

A convenient abbreviation of the model is TM(F,$\{\delta_i(.)\}$). If all the δ-functions have the same structure, instead of explicitly giving the functions we will use, for example, TM(normal, linear) if $F(.)$ is the (standard) normal distribution function, and the difficulty is linear, i.e., of the form $\delta_i(y) = \delta_{0i} + \delta_i y$.

Functions that are of special interest for the characterization of the model are the item characteristic functions and the person threshold functions. The item characteristic functions are generalizations of the item characteristic functions that are familiar from binary models. They are defined for fixed value y as functions of the person parameter.

Item Characteristic Functions

$$IC_{i,y}(\theta_p) = P(Y_{pi} > y|\theta_p, \alpha_i, \delta_i(.)) = F(\alpha_i(\theta_p - \delta_i(y))), \quad y \in S. \quad (6.2)$$

For each $y \in S$ one has one item characteristic (IC) function. It shows the probability for an response above a fixed value y for varying person parameters. In contrast to

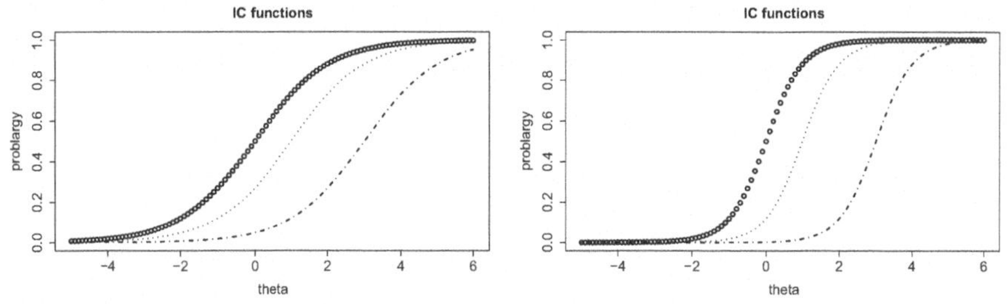

Fig. 6.1 Item characteristic curves for $\delta(y) = y$ for $y = 0$, $y = 1$, $y = 3$, left: $\alpha_i = 1$, right: $\alpha_i = 2$

binary models where only the value $y = 0$ is interesting, for responses with more than two possible outcomes one has several item characteristic functions. If the response is continuous any value y can occur. The functions depend on the item i *and* y.

The shapes of the response functions are determined by the function $F(.)$. All response functions are strictly increasing in θ_p indicating that the probability of scoring larger than y increases with the person parameter. Figure 6.1 shows the item characteristic curves for the simple difficulty function $\delta_i(y) = y$ for values $y = 0$, $y = 1$, $y = 3$ and logistic function $F(.)$. In the left panel $\alpha_i = 1$ has been used, in the right panel $\alpha_i = 2$. Curves for fixed items and different values of y do not cross since $P(Y_{pi} > y_1|\theta_p, \alpha_i, \delta_i(.)) \geq P(Y_{pi} > y_2|\theta_p, \alpha_i, \delta_i(.))$ if $y_1 < y_2$.

Another set of interesting functions are the *person threshold functions* (PT functions), which are defined for fixed item and fixed person parameter θ_p as functions of y.

Person Threshold Functions

$$g_{i,\theta_p}(y) = P(Y_{pi} > y|\theta_p, \alpha_i, \delta_i(.)) = F(\alpha_i(\theta_p - \delta_i(y))) \quad (6.3)$$

Person threshold (PT) functions show the probability of an response above y for a specific person with ability θ_p. They are decreasing functions since the probability of a response above y has to decrease with increasing y. Person threshold functions are in particular interesting if responses can take many different values. For illustration Fig. 6.2 shows on the left-hand side PT functions for three values of θ_p and simple difficulty function $\delta(y) = y$. It is seen that a person with a larger ability has a larger probability of an response above y for all values y than a person with a lower ability. On the right-hand side the corresponding densities are given if the response variable is assumed to be continuous. It is seen that persons with larger ability tend to generate larger responses.

PT functions are strongly related to the distribution function of Y_{pi}, which is simply given by $F_{pi}(y) = 1 - F(\alpha_i(\theta_p - \delta_i(y)))$. The distribution function is denoted with the subscript pi to distinguish it from the response function $F(.)$ (which itself is a distribution function). The information in the distribution function of Y_{pi} is the same as in the person threshold function since for continuous responses the density of Y_{pi}

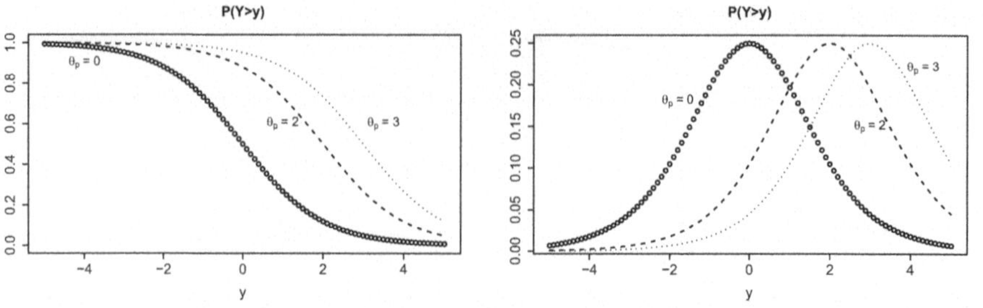

Fig. 6.2 Left: Person threshold functions for $\delta(y) = y$ and logistic response function for three values of θ_p ($\alpha_i = 1$); right: corresponding density

is given by $f_{pi}(y) = \partial F_{pi}(y)/\partial y$. The person threshold function has the advantage that it shows immediately the probability of a response above values y. In item response modeling the focus is on the performance of individuals, and in most tests good performance means high values. Thus it seems natural to investigate if a fixed person is able to score above a value y.

Person thresholds functions reflect the abilities. In general, if $\theta_1 > \theta_2$ the person function generated by θ_1 has larger values than the one generated by θ_2, that is, for all y the inequality $g_{i,\theta_1}(y) \geq g_{i,\theta_2}(y)$ holds.

The third important function in the thresholds model is the *difficulty function* which is evaluated at values from the support S. It is a non-decreasing item-specific function and is defined on the support. While the response function determines the item characteristics curves (probability of scoring larger than y), which always have the shape of $F(.)$ as shown in Fig. 6.1, the difficulty functions crucially determine the distribution of responses. They are essential tools to restrict the domain of the responses, for example, to constrain them to be positive or within a specific interval.

Difficulty functions can be defined on the support as a mapping $\delta_i : \mathbb{R} \longmapsto \mathbb{R} \cup \{\infty, -\infty\}$. Often it is more convenient to use functions that are defined on a wider set, for example, the set of real unumbers. The model then uses only values $y \in S_0$, where $S_0 = S$ if S is infinite, and $S_0 = \{m_1, \ldots, m_{k-1}\}$ for finite support $S = \{m_1, \ldots, m_k\}$. If the support is finite the value of $\delta_i(m_k)$ is implicitly fixed by $\delta_i(m_k) = \infty$. If, for example, responses are binary, $y \in \{0, 1\}$, only the value $\delta_i(0)$ is needed since $P(Y_{pi} > 0) = P(Y_{pi} = 1)$, and $P(Y_{pi} = 0)$ follows from $P(Y_{pi} = 0) = 1 - P(Y_{pi} = 1)$.

In the following we first consider models with a *fixed difficulty function*. The difficulty functions that are considered have the general linear form

$$\delta_i(y) = \delta_{0i} + \delta_i g(y),$$

where $g(.)$ is a strictly increasing function and $\delta_i \geq 0$. The corresponding threshold model with response function $F(.)$ will be abbreviated by TM(F,g). In the case where $g(y) = y$ we will also use TM(F, linear).

There are several ways to make the parameters of the thresholds model identifiable (Tutz, 2022b). Person parameters and item functions are identifiable if one discrimination parameter and one person parameter are fixed, for example, $\alpha_1 = 1, \theta_1 = 0$. An alternative is person parameters and item functions are identifiable if one discrimination parameter is fixed, for example $\alpha_1 = 1$, and $\delta_i(y_0)$ is fixed for one value $y_0 \in S_0$, where $S_0 = S$ if S is infinite and $S_0 = \{m_1, \ldots, m_{k-1}\}$ for finite support $S = \{m_1, \ldots, m_k\}$. In the latter case $\delta_i(m_k) = \infty$ holds for all items.

Binary and Ordinal Models as Special Cases

The binary models considered in previous sections and one specific ordinal model, the graded response model, are special cases of the thresholds model.

The support of binary responses is $S = \{0, 1\}$. Let the difficulty function be defined by $\delta_i(0) = \delta_{i0}$ and $\delta_i(1) = \infty$. Then the corresponding threshold model is

$$P(Y_{pi} = 1|\theta_p, \alpha_i, \delta_i(.)) = P(Y_{pi} > 0|\theta_p, \alpha_i, \delta_i(.)) = F(\alpha_i(\theta_p - \delta_{i0})),$$

which is the homogeneous binary model with response function $F(.)$. Thus, all binary models including the Rasch model are special cases.

Let now the response be polychotomous with support $S = \{0, 1, \ldots, k\}$. Let the difficulty functions be given by $\delta_i(r) = \delta_{i,r+1}$ for $r = 1, \ldots, k-1$, and $\delta_i(k) = \infty$, where the parameters are ordered, that is, $\delta_i(r) \leq \delta_i(r+1)$ for all r. Then, the thresholds model has the form

$$P(Y_{pi} \geq r|\theta_p, \alpha_i, \delta_i(.)) = P(Y_{pi} > r-1|\theta_p, \alpha_i, \delta_i(.)) = F(\alpha_i(\theta_p - \delta_{ir})), \quad r = 1, \ldots, k-1,$$

which is the graded response model (4.4). Thus, one of the general purpose ordinal models turns out to be a special case.

Construction from Binary Models

In Fig. 1.2 the thresholds model has been included as one of the models that can be generated from binary models. In the following it is demonstrated that the threshold model contains binary models but also that the model can be constructed from binary models that serve as building blocks.

If the thresholds model holds one has for a fixed threshold $y_0 \in S$, $P(Y_{pi} > y_0|\theta_p, \delta_i(.)) = F(\alpha_i(\theta_p - \delta_i(y_0)))$. Let a binary response be defined by $Y_{pi}(y_0) = 1$ if $Y_{pi} > y_0$ and $Y_{pi}(y_0) = 0$ otherwise. For this variable the binary item response model

$$P(Y_{pi}(y_0) = 1|\theta_p, \gamma) = F(\alpha_i(\theta_p - \gamma)), \tag{6.4}$$

where $\gamma = \delta_i(y_0)$, holds. Thus, the collapsed binary version $Y_{pi}(y_0)$ of Y_{pi} follows a familiar binary model. That means thresholds models contain binary models. Dichotomization of the response yields models that measure the same latent trait.

The interesting point is that the model can also be constructed from binary models. Let us assume that for any y_0 from the support S a binary model

$$P(Y_{pi}(y_0) = 1|\theta_p, \gamma_{y_0}) = F(\alpha_i(\theta_p - \gamma_{y_0})) \tag{6.5}$$

holds, where $Y_{pi}(y_0)$ denotes the binary response defined previously. Then the thresholds model with the difficulty function defined by $\delta_i(y_0) = \gamma_{y_0}$ holds. Of course, the binary models are strongly linked since they share the parameters α_i, θ_p, and it is assumed that they hold simultaneously.

It means that the thresholds model can also be understood as a collection of binary models that have to hold simultaneously for fixed α_i, θ_p. In this sense, it can be seen as an extension of binary models to models in which the responses have support S. In particular, if $\alpha_i = 1$ and $F(.)$ is the logistic distribution the thresholds model is a collection of Rasch models.

To avoid inconsistencies it should be noted that for finite S the model for $Y_{pi}(y_{\max})$, where $y_{\max}$ is the maximal value of S, is trivial, that is, $P(Y_{pi}(y_{\max}) = 1|\theta_p, \gamma_{y_0}) = 0$ (equivalent to $\gamma_{y_0} = \infty$).

6.2 Linear Difficulty Functions for Continuous Responses

Let the response be continuous with support $\mathbb{R}$. Then an interesting difficulty that is simple but yields already very flexible models is the linear difficulty function. In fact, *any* continuous distribution of Y_p can be generated within the thresholds model framework by using linear difficulty functions. Therefore, in the following let the difficulty function of item i have the form

$$\delta_i(y) = \delta_{0i} + \delta_i y, \quad \delta_i > 0,$$

which yields a linear thresholds model, TM(F, linear). The parameterization makes the model a *parametric model* with item parameters $\delta_{0i}, \delta_i, i = 1, \dots, I$. Let $F(.)$ denote a fixed, typically standardized, distribution function with support $\mathbb{R}$, for example, the standardized normal or the logistic distribution function. Linear difficulty functions yield a simple representation of means and variances of responses. They are given by

$$\mu_{pi} = \mathrm{E}(Y_{pi}) = \gamma_i\theta_p - \gamma_{0i}, \tag{6.6}$$

$$\sigma_{pi}^2 = \mathrm{var}(Y_{pi}) = \frac{\sigma_F^2}{\alpha_i^2\delta_i^2}, \tag{6.7}$$

where $\gamma_i = 1/\delta_i$, $\gamma_{0i} = (\delta_{0i} + \mu_F/\alpha_i)/\delta_i$, and μ_F, σ_F^2 are constants that are determined by the distribution function $F(.)$. More concrete, $\mu_F = \int yf(y)dy$ is the expectation corresponding to distribution function $F(.)$ and $\sigma_F^2 = \mathrm{var}_F = \int(y - \mu_F)^2 f(y)dy$ the corresponding variance. For a derivation see Tutz (2022b).

It is obvious that large values of θ_p indicate a tendency to high responses. The item parameter γ_i is a scaling parameter and γ_{0i} is the location of the item on the latent scale. It represents the "basic" difficulty of the item; if γ_{0i} is large, the expected response tends to be small, and vice versa.

If $F(.)$ is symmetric, $\mu_F = 0$ holds and one obtains the simpler form

$$\mu_{pi} = \mathrm{E}(Y_{pi}) = (\theta_p - \delta_{0i})/\delta_i.$$

Then, the expectation does not depend on the discrimination parameter. However, the discrimination parameter still determines the variance such that large values of the discrimination parameter are associated with small variances of the response.

Homogeneous Linear Thresholds Model

If δ_i and α_i do not depend on the item, that is, $\delta_i = \delta, \alpha_i = \alpha$ for all items one obtains the simpler forms $\mathrm{E}(Y_{pi}) = \theta_p/\delta - \gamma_{0i}$, $\mathrm{var}(Y_{pi}) = \sigma_F^2/\alpha^2\delta^2$, which means that the response is simply determined by the difference between the (scaled) ability θ_p/δ and item difficulty γ_{0i}, a property that is familiar from the binary Rasch model or the normal-ogive model without a slope parameter. Moreover, the variance of the responses is the same for all items and persons. It means that difficulty functions are parallel ($\delta_i = \delta$) and the intercepts of the difficulty functions represent the difficulties of items (up to a constant). The thresholds model with fixed slopes and discrimination parameters, $\delta_i = \delta$, $\alpha_i = \alpha$, is referred to as *homogeneous linear thresholds model*, or shortly *homogeneous* TM(F, linear).

The left column of Fig. 6.3 illustrates the homogeneous model with parameters $(2, 1), (0, 1), (-2, 1)$ (for (δ_{0i}, δ_i)) and $\alpha_i = 1$. The first picture in the top row shows the difficulty functions of the three items. Since the slopes of the difficulty functions are the same for all items ($\delta_i = 1$) difficulty functions are parallel, only intercepts determine the difficulty. Since item 1 has the highest intercept it is the most difficult one (under the assumption that high scores mean high performance). The second picture (left column) shows the corresponding densities for $\theta_p = 0$ if $F(.)$ is the normal distribution. The smallest responses are to be expected for item 1, for which the difficulty function has the largest values. The third picture (left column) shows the densities for a non-symmetric response function $F(.)$, namely the Gompertz distribution $F(y) = 1 - \exp(-\exp(y))$. The important feature seen from the left column of Fig. 6.3 is that densities of item responses are simply shifted versions of the same distribution with the shifting determined by the intercepts of the difficulty function.

The picture changes in the non-homogeneous case. The second column of Fig. 6.3 shows difficulty functions and densities if slopes are varying. Parameters are given by $(2, 0.8), (0, 1), (-2, 1.2)$ (for (δ_{0i}, δ_i)), $\alpha_i = 1$. Then difficulty functions are not parallel and the difficulty of the items is determined by the intercepts *and* the slopes. The densities are no longer shifted versions, they are shifted and scaled versions. For example, item 3 has the smallest variance since it has the highest slope (second and third picture, right column).

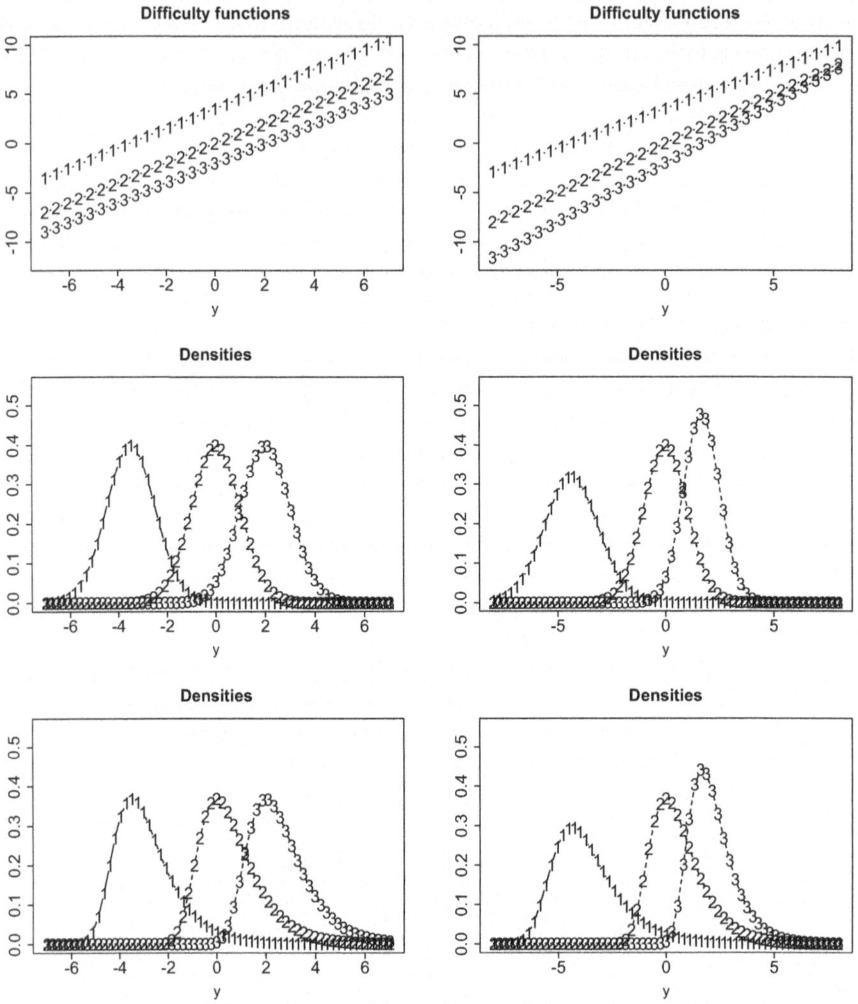

Fig. 6.3 Difficulty functions and densities for three items with equal slopes and varying intercepts (first column), and for three items with varying slopes (second column). Second row: normal response function, third row: Gompertz for $F(.)$

Distribution of Responses

In the following, the distributions of responses are investigated in more detail. If the response function $F(.)$ is *symmetric*, as for example the logistic distribution, one obtains that the distributions of the responses are simply scaled and shifted versions of the distribution $F(.)$. More concise, let $f(y) = \partial F(y)/\partial y$ denote the density linked to the response function $F(.)$, and $f_{pi}(.)$ and $F_{pi}(.)$ denote the density and distribution function of the response Y_{pi}, respectively. Then one obtains

$$f_{pi}(y) = \partial F_{pi}(y)/\partial y = f(\alpha_i(\delta_{0i} + \delta_i y - \theta_p))\alpha_i\delta_i,$$
$$F_{pi}(y) = P(Y_{pi} \le y) = F(\alpha_i(\delta_{0i} + \delta_i y - \theta_p)). \tag{6.8}$$

If the response function $F(.)$ is *not symmetric* the form of the distribution function of the responses is not the same as for distribution function $F(.)$. Instead, the distribution of Y_{pi} is given as a scaled and shifted version of the distribution function $\bar{F}(y) = 1 - F(-y)$. In general, if a random variable Y has distribution function $F(.)$, the random variable $-Y$ has distribution function $\bar{F}(.)$. If $F(.)$ is symmetric, there is no difference, $\bar{F}(y) = F(y)$. With $\bar{f}(y) = \partial\bar{F}(y)/\partial y$ denoting the density corresponding to $\bar{F}(.)$ densities and distribution functions are given by

$$f_{pi}(y) = \partial F_{pi}(y)/\partial y = \bar{f}(\alpha_i(\delta_{0i} + \delta_i y - \theta_p))\alpha_i\delta_i = f(\alpha_i(\theta_p - \delta_{0i} - \delta_i y))\alpha_i\delta_i,$$
$$F_{pi}(y) = P(Y_{pi} \le y) = \bar{F}(\alpha_i(\delta_{0i} + \delta_i y - \theta_p)) = 1 - F(\alpha_i(\theta_p - \delta_{0i} - \delta_i y)), \tag{6.9}$$

which simplifies to (6.8) if $F(.)$ is symmetric. For continuous responses, the response function $F(.)$ determines the distributional form of the responses. The parameters determine the location and the spread. In Fig. 6.3 the densities for a person with $\theta = 0$ are shown for the symmetric normal response function (second row). If for $F(.)$ the Gompertz distribution is assumed, which is given by $F(y) = 1 - \exp(-\exp(y))$, the responses follow a Gumbel distribution, which has distribution function $\bar{F}(y) = \exp(-\exp(-y))$ (third row).

Total Score

A classical measure for the performance of persons is the total score $Y_p = \sum_{i=1}^{I} Y_{pi}$. Conditions under which the total score is an appropriate measure have been investigated in particular for categorical responses, see, for example, Masters (1982); Hemker et al. (1997); Sijtsma and Hemker (2000); Hemker et al. (2001). For the TM with linear difficulty functions one obtains for the mean and variance of the total score

$$\mu_p = \mathrm{E}(Y_p) = \gamma\theta_p - \gamma_0, \quad \sigma_p^2 = \mathrm{var}(Y_p) = \gamma_\sigma, \tag{6.10}$$

where $\gamma = \sum_i^I \gamma_i$, $\gamma_0 = \sum_i^I \gamma_{0i}$, $\gamma_\sigma = \sum_i^I (\sigma_F^2/(\alpha_i^2\delta_i^2))$ and conditional independence has been assumed to hold. Thus, for linear difficulty functions the mean of the total score is a linear function of the latent trait. Since γ, γ_0 are constants that do not depend on single items one could also consider the scaled ability $\tilde{\theta}_p = \gamma\theta_p - \gamma_0$. Then, the expectation of Y_p is equal to the scaled ability.

6.3 Non-linear Difficulty Functions for Continuous Responses

As shown in Sect. 6.2 any continuous distribution with support $\mathbb{R}$ can be generated within the thresholds model framework by using linear difficulty functions. Non-linear difficulty functions are in particular helpful if the response is restricted, for example, to be positive or within a fixed interval. Restrictions of this type are obtained in a natural way by specifying appropriate non-linear difficulty functions. If difficulty functions are non-linear also the mean and other characteristics of the responses are non-linear functions of the latent trait. In the following, it is assumed that difficulty functions have the form

$$\delta_i(y) = \delta_{0i} + \delta_i g(y),$$

where $g(.)$ is a strictly increasing fixed function. The corresponding parametric thresholds model is denoted by TM(F,g(.)).

Responses in the Positive Domain

The threshold model automatically restricts the responses to positive values, if for the difficulty function $\lim_{y \to 0} \delta_i(y) = -\infty$ holds. One candidate is the logarithmic function $g(y) = \log(y)$ yielding $\delta_i(y) = \delta_{0i} + \delta_i \log(y)$.

For illustration Fig. 6.4 shows logarithmic difficulty functions for three items with (δ_{0i}, δ_i) given by $(-3, 2), (-3.8, 2), (-6, 2)$, $(\alpha_i = 1)$. Figure 6.5 shows the corresponding response distributions if $\theta_p = 0$. The left picture shows the distribution if the response function is the normal distribution, on the right side the Gompertz distribution has been used as response function. The difficulty function of item 3 has smaller values than the difficulty functions of the other items. Consequently, for item 3 larger responses are to be expected than for other items.

It is seen that for non-linear difficulty functions the density is not a shifted and scaled version of the density linked to $F(.)$. The difficulty function transforms the density in a non-linear way. With $F_{pi}(.)$ denoting the distribution function of Y_{pi} and $f_{pi}(y) = \partial F_{pi}/\partial y$, the corresponding density for continuous responses one obtains from $F_{pi}(y) = P(Y_{pi} \leq y) = 1 - F(\alpha_i(\theta_p - \delta_i(y)))$.

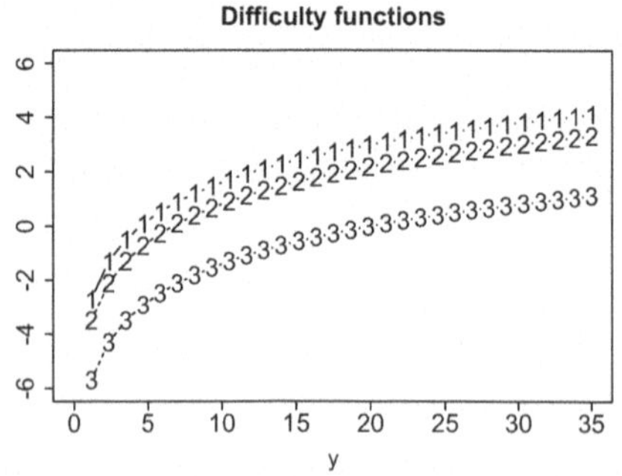

Fig. 6.4 Logarithmic difficulty functions for three items with intercepts and slopes given by $(-3, 2), (-3.8, 2), (-6, 2)$

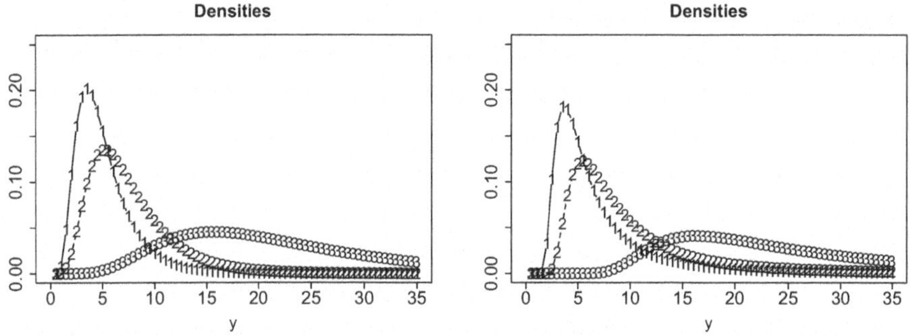

Fig. 6.5 Densities for three items, logarithmic difficulty functions, left: normal response function with intercepts and slopes given by (−3, 2), (−3.8, 2), (−6, 2), right: Gumbel for $\bar{F}(.)$

$$f_{pi}(y) = f(\alpha_i(\theta_p - \delta_i(y)))\alpha_i\delta_i'(y),$$

where $f(.)$ is the density corresponding to $F(.)$, and $\delta_i'(y) = \partial\delta_i(y)/\partial y$ is the derivative of the threshold function.

Means and variances of responses cannot be expected to be linear functions of the person parameter as in the case of linear difficulty functions. For strictly increasing difficulty functions one obtains the general result

$$\mathrm{E}(Y_{pi}) = \int_0^1 \delta_i^{-1}(\theta_p - F^{-1}(1-q)/\alpha_i)dq, \tag{6.11}$$

$$\mathrm{E}(Y_{pi} - \mu_{pi})^k = \int_0^1 (\delta_i^{-1}(\theta_p - F^{-1}(1-q)/\alpha_i) - \mathrm{E}(Y_{pi}))^k dq \tag{6.12}$$

(Tutz and Jordan, 2024). One consequence is that the total score $Y_p = \sum_i Y_{pi}$, which is often used to measure the ability, is not appropriate.

Much simpler formulae are obtained if one considers the transformed responses $\delta_i(Y_{pi})$ instead of the means and variances of responses themselves. One can derive

$$\mathrm{E}(\delta_i(Y_{pi})) = \theta_p - \mu_F/\alpha_i, \quad \mathrm{var}(\delta_i(Y_{pi})) = \sigma_F^2/\alpha_i^2, \tag{6.13}$$

where μ_F, var_F are again the expectation and variance corresponding to distribution function $F(.)$ (Tutz and Jordan, 2024). Then, one obtains for the expected *transformed total score* $Y_p^{(\delta)} = \sum_i \delta_i(Y_{pi})$ a linear function of the ability,

$$\mathrm{E}(Y_p^{(\delta)}) = I\theta_p - \sum_{i=1}^{I} \mu_F/\alpha_i,$$

see also Sect. 6.4.

Logarithmic Difficulty Function

Simpler expressions can be obtained in special cases as the logarithmic difficulty function, which is considered in the following. If $\delta_i(y) = \delta_{0i} + \delta_i \log(y)$ one obtains

$$E(Y_{pi}) = c_i \exp\left(\frac{\theta_p - \delta_{0i}}{\delta_i}\right),$$

where c_i is a constant that depends on α_i, δ_i ($c_i = \int_0^1 \exp(-F^{-1}(1-q)/(\alpha_i\delta_i))dq$). Thus, expectations are exponential functions of the latent ability.

Then, the total score is given by

$$\mathrm{E}(Y_p) = \sum_{i=1}^{I} c_i \exp\left(\frac{\theta_p - \delta_{0i}}{\delta_i}\right) = \sum_{i=1}^{I} c_i \exp(-\delta_{0i}/\delta_i)\exp(\theta_p)^{1/\delta_i},$$

which is a weighted sum of exponential terms with the terms depending on the item. Only in special cases the total score can be considered as a measurement of abilities. If items are homogeneous, that is, $\delta_i = \delta$ for all i, one obtains

$$\mathrm{E}(Y_p) = \sum_{i=1}^{I} c_i \exp(-\delta_{0i}/\delta)\exp(\theta_p/\delta) = c\exp(\tilde{\theta}_p),$$

where $\tilde{\theta}_p = \theta_p/\delta$ is the scaled ability and $c = \sum_i c_i \exp(-\delta_{0i}/\delta)$. Then, $\mathrm{E}(Y_p)$ is an exponential function of the scaled ability $\tilde{\theta}_p$. Of course, one can also consider the transformed ability $\exp(\tilde{\theta}_p)$ as a measure of ability. Then, $\mathrm{E}(Y_p)$ depends linearly on the (transformed) ability. Thus, in the homogeneous case, but only then, the total score can be considered as being directly linked to the underlying ability.

An interesting case is the model TM(F,log) if $F(.)$ is the standard normal distribution. Then, the density of responses becomes

$$\begin{aligned} f_{pi}(y) &= f(\alpha_i(\delta_{0i} + \delta_i \log(y) - \theta_p))\alpha_i\delta_i/y \\ &= (\sqrt{2\pi})^{-1}\exp(-(\alpha_i(\delta_{0i} + \delta_i\log(y) - \theta_p))^2/2)\alpha_i\delta_i/y \\ &= \frac{1}{y\sqrt{2\pi}\bar{\sigma}_i}\exp(-\frac{(\log(y) - \bar{\mu}_{pi})^2}{2\bar{\sigma}_i^2}), \end{aligned}$$

where $\bar{\mu}_{pi} = (\theta_p - \delta_{0i})/\delta_i$, $\bar{\sigma}_i = 1/(\alpha_i\delta_i)$. This is the lognormal distribution with parameters $\bar{\mu}_{pi}, \bar{\sigma}_i$. The homogeneous version of the model ($\delta_i = \delta$) is equivalent to van der Linden's lognormal response-time model, which is a speed model that carefully distinguishes between time and speed (van der Linden, 2016). The thresholds version of van der Linden's model is a generalization allowing for varying slope parameters. It also offers the possibility to consider alternative response functions that replace the normal distribution and might yield better fit.

Responses in an Interval

In some applications responses are restricted to a fixed interval, for example if responses are on a visual analogue scale or if responses are recorded as numbers between 0 and 100. A classical case are Likert items, which typically take numbers between 1 and 5 or 1 and 7 indicating degrees of agreement or disagreement to questions. They are special since they are discrete and it is unclear if responses are interval-scaled. In the following, we consider the case of continuous response distributions. Likert scales will be considered separately in Sect. 6.5.

Let responses be known to be restricted to the interval (a, b). Thresholds models can account for the restriction in a very simple way, namely by choosing a difficulty function for which $\lim_{y \to a} \delta_i(y) = -\infty$ and $\lim_{y \to b} \delta_i(y) = \infty$ hold since then $P(Y_{pi} > a) \to 1$ and $P(Y_{pi} > b) \to 0$. A difficulty function that meets these demands is the logit function $g(y) = \log((y-a)/(b-y))$ yielding $\delta_i(y) = \delta_{0i} + \delta_i \log((y-a)/(b-y))$. For simplicity one can also transform the data into the interval $(0, 1)$ and use the simpler function $g(y) = \log(y/(1-y))$. Alternatively, any distribution function can be used in the difficulty function. Let F_{diff} be a strictly increasing distribution function. Then, $g(y) = F_{\text{diff}}^{-1}(.)$ restricts the responses to the interval $(0, 1)$. The logit function results if F_{diff} is chosen as the logistic distribution function.

For illustration Fig. 6.6 shows the distributions that are obtained for the interval $(0, 1)$ and standard normal distribution $F(.)$. The underlying item parameters are $(3.3, 2), (1, 2), (-2.5, 2)$ (for (δ_{0i}, δ_i)). The left picture shows the distribution of responses if $\theta_p = 0$, in the right picture $\theta_p = 1.5$. It is seen that responses are within the interval $(0, 1)$. For larger values of θ_p the distribution is not just shifted but distinctly changes its form.

The expectations and variances of responses are again given by (6.11). As in the case of logarithmic difficulty functions they depend on the person's ability θ_p. Also the simple total score depends on the ability in a complex form and cannot be considered an appropriate measure of the underlying latent trait. However, one can

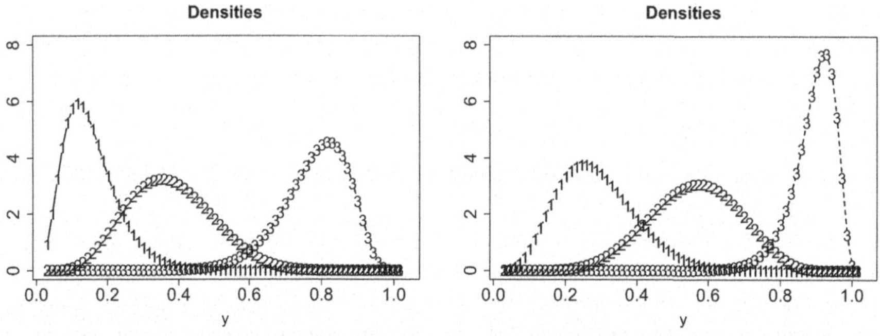

Fig. 6.6 Distributions of responses for items with intercept, slope given by $(3.3, 2), (1, 2), (-2.5, 2)$ with logit difficulty function, left: $\theta = 0$, right: $\theta = 1.5$

again use the *transformed total score* $Y_p^{(\delta)} = \sum_i \delta_i(Y_{pi})$, for which the general result given in Eq. (6.13),

$$\mathrm{E}(Y_p^{(\delta)}) = I\theta_p - \sum_i \mu_F/\alpha_i,$$

holds. The transformed total score is quite different from the traditional total score since the transformation is highly non-linear. It is much more appropriate as an indicator of the underlying ability than the traditional score given by sums of responses.

The case of responses in intervals is especially important in Likert-type items, which by definition are restricted to a fixed interval $[1, m]$ in m-grade Likert scales. Ignoring the restriction to an interval yields improper densities, which typically are positive beyond the interval $[1, m]$. In addition, the total score as a sum of responses over items is not a reliable indicator of the latent trait. It will be considered in more detail in the next section.

The thresholds model for intervals of the form (a, b) is equivalent to Samejima's homogeneous continuous response model (Samejima, 1973). Alternative models for responses that allow responses in intervals are the beta-binomial model considered by Noel and Dauvier (2007) and the zero and one inflated item response theory modeling framework proposed by Molenaar et al. (2022). The latter allows responses on the boundaries by combining the discrete graded response model with continuous models for the interval (a, b). An overview on the modeling of visual analog scales has been given by Heller et al. (2016), the inclusion of response styles has been considered by Minusa et al. (2024). A program that fits regression models for the visual analog scale has been provided by Manuguerra et al. (2020).

Example 6.1 Self-regulation (Tutz and Jordan, 2024)
The data set *Lakes* from the R package *MPsychoR* (Mair, 2018) is a multi-facet G-theory application taken from Lakes and Hoyt (2009). The authors used the response to assess children's self-regulation in response to a physically challenging situation. The scale consists of three domains, cognitive, affective/motivational, and physical. We use the physical domain only. Each of the 194 children was rated by 5 raters on six items on his/her self-regulatory ability with ratings on a scale from 1 to 7. We use the average rating over the five raters, which yields a response that takes values in the interval $(1, 7)$ but is not confined to integer values.

Table 6.1 shows the obtained marginal log-likelihoods and estimates of standard deviation of the mixing distribution σ_θ for several response and difficulty functions with varying slopes in the difficulty functions. The columns on the left show fits for fixed discrimination parameters ($\alpha_i = 1$), the right columns show fits for varying discrimination parameters. It is seen that varying discrimination parameters yield significantly better fits (last column shows likelihood ratio tests). For example, the likelihood ratio test that compares the fixed discrimination model and the varying discrimination model is 21.090 on 2 df for the Gumbel response function with logit difficulty function. Similar results hold for the other models. For given response function F, the best fits are always obtained for logit difficulty functions. This is to be expected since the linear and logistic difficulty functions ignore the restriction

Table 6.1 Fits for self-regulation data

		$\alpha_i = 1$		α_i varying		Lik-ratio test
Response fct	Difficulty fct	Log-lik	$\hat{\sigma}_{\theta_p}$	Log-lik	$\hat{\sigma}_{\theta_p}$	
NV	Linear	−563.084	1.493	−554.421	2.596	17.326
	Log	−654.136	1.260	−644.621	2.379	19.030
	Logit	−521.785	1.729	−515.111	2.264	13.348
logistic	Linear	−542.951	2.753	−531.891	4.404	22.120
	Log	−594.417	2.507	−561.602	6.370	65.630
	Logit	−509.584	3.179	−500.133	4.324	18.902
Gumbel	Linear	−525.132	1.758	−514.559	2.643	21.149
	Log	−552.629	1.584	−549.302	2.607	6.654
	Logit	−508.218*	1.989	−497.673*	2.948	21.090

to the interval, and have only been included to demonstrate that they yield poor fits. For a given difficulty function, the best fit has been obtained by choosing the Gumbel distribution as response function F. The Gompertz response function has been omitted since the fits were poor, for more fits, see Tutz and Jordan (2024).

Figure 6.7 shows the estimated response densities for the three items with varying discrimination parameters (using as response function the Gumbel distribution). First row shows linear difficulty functions, second row logit difficulty functions, left column shows responses for latent trait $\theta_{\text{low}} = 0$, right column shows responses for latent trait $\theta_{\text{high}} = 3.5$, which is not extreme given $\hat{\sigma}_{\theta_p}$ is larger than 2.5. It is seen that the linear difficulty function yields inappropriate densities that take values outside the interval $(1, 7)$. The logit difficulty function, which restricts the responses, yields much more appropriate distributions. □

Responses in a Closed Interval

If responses are in a closed interval $[a, b]$ with boundary points a and b having positive probabilities, the response distribution is not continuous but a combination of continuous and discrete distributions. Models for this type of response are the inflation models considered by Molenaar et al. (2022). They refer to it as zero and one inflated bounded IRT models since they considered the interval [0, 1]. A thresholds model constructed in the spirit of the inflation model for interval $[a, b]$ is the model

$$
\begin{aligned}
&P(Y_{pi} = a|\theta_p, \tau_i) = 1 - F_o(\zeta_i(\theta_p - \gamma_{i0})),\\
&P(Y_{pi} > y|\theta_p, \tau_i) = (F_o(\zeta_i(\theta_p - \gamma_{i1})) - F_o(\zeta_i(\theta_p - \gamma_{i0})))F(\alpha_i(\theta_p - \delta_i(y))), \text{ for } y \in (a, b)\\
&P(Y_{pi} = b|\theta_p, \tau_i) = F_o(\zeta_i(\theta_p - \gamma_{i1})),
\end{aligned}
$$

where $F_o(.)$ and $F(.)$ are strictly monotone distribution functions and $\delta_i(y)$ is a difficulty function that ensures that responses are in (a, b), for example, $\delta_i(y) = \delta_{0i} +$

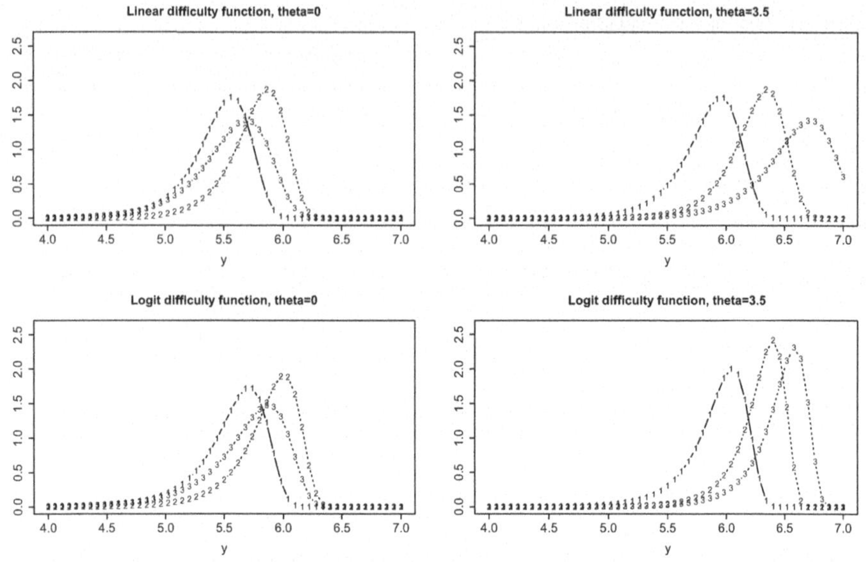

Fig. 6.7 Estimated densities for self-regulation, first row: linear difficulty functions, second row: logit difficulty functions, left: $\theta_{\text{low}} = 0$, right: $\theta_{\text{high}} = 1.5$

$\delta_i \log(y - a)/(b - y)$. The item parameters $\alpha_i, \zeta_i, \gamma_{i0}, \delta_i(.)$ are collected in τ_i. The underlying structure of the model becomes clear when considering the categorical responses $\bar{Y}_{pi} = 0$ if $Y_{pi} = a$, $\bar{Y}_{pi} = 1$ if $Y_{pi} \in (a, b)$, $\bar{Y}_{pi} = 2$ if $Y_{pi} = b$. For the categorical response $\bar{Y}_{pi}$ the ordinal graded response model

$$P(\bar{Y}_{pi} \geq r|\theta_p, \tau_i) = F_o(\zeta_i(\theta_p - \gamma_{ir})), \quad r = 0, 1,$$

holds with response function $F_o(.)$, discrimination parameter ζ_i and location parameters γ_{io}, γ_{i1}. The thresholds model for continuous responses with response function $F(.)$ and difficulty functions $\delta_i(.)$ is embedded coming into play for responses in the interval (a, b).

The model explicitly allows for positive probabilities at end points. Molenaar et al. (2022) used alternative parametric models for the conditional response given $Y_{pi} \in (a, b)$. An advantage of the thresholds model is that it is very flexible and quite different distributions can be modeled, in particular if one uses the data-adaptive approach that uses basis functions as outlined in Sect. 6.6.

6.4 Further Properties of Thresholds Models

Transformed Responses

At the heart of thresholds models for continuous responses there is a linear relationship between abilities and expectations, however it does not relate to the expectations of the responses itself but to transformed responses. More concise, let $F(.)$ be strictly increasing, Y_{pi} have continuous distribution, and item difficulties have the form $\delta_i(y) = \delta_{0i} + \delta_i g(y)$. Then, for item i the mean and variance of the *transformed* response $\delta_i(Y_{pi})$ are given by

$$\mathrm{E}(\delta_i(Y_{pi})) = \theta_p - \mu_F/\alpha_i, \quad \mathrm{var}(\delta_i(Y_{pi})) = \sigma_F^2/\alpha_i^2, \tag{6.14}$$

where μ_F, var_F are again the expectation and variance corresponding to distribution function $F(.)$ (Tutz and Jordan, 2024).

One consequence is that the expected *transformed total score* $Y_p^{(\delta)} = \sum_i \delta_i(Y_{pi})$ is a linear function of the ability,

$$\mathrm{E}(Y_p^{(\delta)}) = I\theta_p - \sum_{i=1}^{I} \mu_F/\alpha_i.$$

Thus, $Y_p^{(\delta)}$ can be seen as an indicator of the ability. It has the form

$$Y_p^{(\delta)} = \sum_{i=1}^{I} \delta_{0i} + \sum_{i=1}^{I} \delta_i g(Y_{pi}),$$

which is a weighted sum of transformed responses.

For fixed difficulty function a simple general relationship holds between the TM and the chosen response function $F(.)$. It can be shown that

$$\delta_i(Y_{pi}) \text{ has the same distribution as } \theta_p - Y_0/\alpha_i,$$

where Y_0 follows the distribution function $F(.)$ Hence, on the level of the transformed variable, the latent ability θ acts as a simple location parameter, thereby shifting expectations and quantiles in a linear manner.

TM model for Transformed Responses

While the transformation $\delta_i(Y_{pi})$ uses an item-specific transformation, $g(Y_{pi})$ uses the function $g(.)$ only and can be computed without reference to specific items. Therefore, one can consider working with the transformed responses $g(Y_{pi})$ and formulate models for the transformed responses. The corresponding mean and variance are given by

$$E(g(Y_{pi})) = \frac{\theta_p - \mu_F/\alpha_i - \delta_{i0}}{\delta_i}, \quad \text{var}(g(Y_{pi})) = \frac{\sigma_F^2}{\alpha_i^2\delta_i^2}.$$

Let us consider the thresholds model with transformation $g(.)$, TM(F, $g(y)$), which is given by

$$P(Y_{pi} > y|\theta_p, \alpha_i, \delta_i(.)) = F(\alpha_i(\theta_p - \delta_{0i} - \delta_i g(y))).$$

If the model holds one obtains for the transformed responses $g(Y_{pi})$

$$P(g(Y_{pi}) > z|\theta_p, \alpha_i, \delta_i(.)) = P(Y_{pi} > g^{-1}(z)|\theta_p, \alpha_i, \delta_i(.)) = F(\alpha_i(\theta_p - \delta_{0i} - \delta_i z)),$$

which means that $g(Y_{pi})$ follows the *linear* threshold model TM(F, linear) with the same item parameters. Thus, one can alternatively fit the linear model for the transformed responses $g(Y_{pi})$, which yields the same parameter estimates as the original model.

However, some care is needed when comparing alternative models since the log-likelihoods of TM($F, g(.)$) for the original data and TM(F, linear) for the transformed data $g(Y_{pi})$ are not the same. As an example we use the self-regulation data. The log-likelihood obtained when fitting a model with normal response function and logarithmic difficulty function is −654.136. When fitting a model with linear difficulty function to the log-transformed data one obtains 326.531. These models should definitely not be compared via goodness-of-fit measures based on their log-likelihoods although parameter estimates for both models are identical.

A special model that uses transformed responses has been proposed by Wang and Zeng (1998). They transform data using a logistic transformation $\log(y) = y/(m - y)$, where m is the highest score to obtain responses in the interval (0, 1) from positive-valued continuous responses. The approach is similar but the transformation within the framework of thresholds models is more general and refers to the difficulty function, in which an intercept is included.

Independence of Measurement Instruments

Thresholds models generally show independence of measurement instruments (separability of parameters) if the discrimination parameters are fixed ($\alpha_i = 1$). Then one has for two persons p_1, p_2 and any $y \in S$

$$F^{-1}(P(Y_{p_1 i} > y|\theta_{p_1}, \delta_i(.))) - F^{-1}(P(Y_{p_2 i} > y|\theta_{p_2}, \delta_i(.))) = \theta_{p_1} - \theta_{p_1}, \quad (6.15)$$

which does not depend on the item. Thus persons can be compared independently of the item involved. In fact, there is not only one way to compare persons but various ways if the support S has more than two values since Eq. (6.15) holds for any $y \in S$.

For two items i_1, i_2 one has

$$F^{-1}(P(Y_{pi_1} > y|\theta_p, \delta_{i_1}(.))) - F^{-1}(P(Y_{pi_2} > y|\theta_p, \delta_{i_2}(.))) = -(\delta_{i_1}(y) - \delta_{i_2}(y)),$$

which does not depend on the person. Since it holds for any $y \in S$ differences of the whole difficulty functions can be obtained.

The independence of the measurement instrument is a consequence of the independence found in binary models with $\alpha_i = 1$ and the strong link between thresholds models and binary models.

6.5 Likert Scales: Discrete or Continuous?

It has been extensively debated if Likert scales can be regarded as ordinal or interval scales, see, for example, Carifio and Perla (2007), Jamieson (2004), Harpe (2015). Lantz (2013) explicitly investigated the assumption of equidistance of Likert-type scales and used an experimental design to show that the perceived distance between scale points on a regular five-point Likert-type scale depends on how the verbal anchors are used. Harpe (2015) gave an overview concerning problems and the pros and cons of modeling Likert scales modeled as continuous responses. He distinguished between the "ordinalist" and the "intervalist" view. The assumption of the scale level is crucial since it determines the tools that can be used to analyze data, for example, the mean and standard deviation are inappropriate for ordinal data, also the use of item sum then seems doubtful. If Likert scales are designed to measure an underlying latent variable whose values characterizes the respondents' attitudes it is certainly safer to assume not more than ordinal scale level for the items.

The discussion has somehow overshadowed another problem, namely if Likert scales, which are by definition discrete, can be analyzed by using continuous response distributions. Of course, in practice all measurements are discrete since the precision of measurement instruments is always limited. However, if the number of possible outcomes is very large, continuous modeling has distinct advantages and should be contemplated. The question is if the number of possible outcomes in Likert scales is large enough that a continuous approximation is warranted. Even more important is that one does not use classical models that assume a normal distribution of responses since they are totally inappropriate if responses close to boundaries are to be expected, which is bound to happen for very small and very large person parameters.

The thresholds model provides a tool to compare discrete and continuous modeling since it allows for both types of distributions, and accounts for the restricted support. The distribution function for observation Y_{pi} has the general form $F_{pi}(y) = P(Y_{pi} \leq y) = 1 - F(\alpha_i(\theta_p - \delta_i(y)))$. If responses are *continuous* the density is given by

$$f_{pi}(y) = \frac{\partial F_{pi}(y)}{\partial y} = f(\alpha_i(\theta_p - \delta_i(y)))\alpha_i \delta_i'(y),$$

where $f(.)$ is the density corresponding to $F(.)$, and $\delta_i'(y) = \partial \delta_i(y)/\partial y$ is the derivative of the threshold function. If responses are *discrete* with $Y_{pi} \in \{1, \ldots, k\}$

responses are multinomially distributed and the probability mass function is obtained by building differences. It is given by

$$f_{pi}(1) = 1 - P(Y_{pi} > 1) = 1 - F(\alpha_i(\theta_p - \delta_i(0))),$$
$$f_{pi}(r) = P(Y_{pi} > r - 1) - P(Y_{pi} > r) = F(\alpha_i(\theta_p - \delta_i(r-1))) - F(\alpha_i(\theta_p - \delta_i(r))),$$
$$r = 2, \ldots, k.$$

The most attractive difficulty functions are the ones that enforce that the response is restricted to the interval $[1, k]$. A candidate to obtain a restriction to the finite interval (a, b) is the logit difficulty function $g(y) = \log((y - a)/(b - y))$. However, when approximating a Likert scale with values $1, \ldots, k$ it is not sensible to define the logit function for the interval $(1, k)$ since it excludes the extreme values 1 and k. To include them the interval has to be enlarged. Therefore, when fitting a continuous response model we use the interval $(1 - c, k + c)$, where c is a constant, for example $c = 0.5$. When fitting a discrete thresholds model, the symmetric widening of the interval is not appropriate because the difficulty function should tend to infinity at the upper boundary. Therefore, in the discrete case we use the interval $(1 - 2c, k)$. Alternatives to the logit function are functions $g(y) = F_{\text{diff}}^{-1}((y - a)/(b - a))$, where F_{diff} is any strictly increasing distribution function. Then, the responses are restricted to the interval (a, b), which have to be widened as in the case of the logit difficulty function if responses are from $[1, k]$, which includes the boundaries.

Example 6.2 Fears data (Tutz and Jordan, 2024)
As an illustrating example, we consider again the fears data from the German Longitudinal Election Study (GLES). The participants were asked: "How afraid are you due to the ..." - (1) refugee crisis? - (2) global climate change? - (3) international terrorism? - (4) globalization? - (5) use of nuclear energy? The answers were measured on Likert scales from 1 (not afraid at all) to 7 (very afraid). The model is fitted under the assumption that fear is the dominating latent trait, which is considered as unidimensional. We use 200 persons sampled randomly from the available set of observations.

Table 6.2 shows the log-likelihoods, the AIC values, and estimates of the standard deviation of the mixing distribution σ_{θ_p} for various responses and difficulty functions. The models have varying slopes in the difficulty functions and fixed discrimination parameters since varying discrimination parameters did not improve the fit significantly. We considered two fits, assuming that the responses are continuous or discrete. It is seen that in both cases the logit difficulty function, defined on $(0.5, 7.5)$ for continuous models and defined on $(0, 7.01)$ for discrete models, always fits better than linear or logarithmic functions. The best fit is obtained when combining logit difficulty functions with a logistic response function, although normal and Gumbel response functions yield rather similar results. The fits of the Gompertz response function were poor and are not shown, but see Tutz and Jordan (2024).

Figure 6.8 shows the fitted person threshold functions for discrete and continuous models (logit difficulty functions, logistic response function). It is seen that the

Table 6.2 Thresholds models for fears data

		Continuous			Discrete		
Response fct	Diff fct	Log-lik	AIC	$\hat{\sigma}_{\theta_p}$	Log-lik	AIC	$\hat{\sigma}_{\theta_p}$
NV	lin	−1847.53	3717.061	0.707	−1847.698	3717.396	0.722
	Log	−2046.301	4114.602	0.719	−2064.723	4151.440	0.758
	Logit	−1703.461	3428.921	0.705	−1735.862	3493.724	0.752
Logistic	lin	−1849.513	3721.026	1.343	−1822.079	3666.158	1.390
	Log	−2000.54	4023.08	1.331	−2024.321	4070.643	1.384
	Logit	−1700.976*	3423.952*	1.327	−1728.913*	3479.826*	1.339
Gumbel	lin	−1817.634	3657.269	0.799	−1819.489	3660.978	0.830
	Log	−1891.852	3805.704	0.629	−1915.859	3853.718	0.619
	Logit	−1704.79	3431.579	0.856	−1734.455	3490.910	0.915

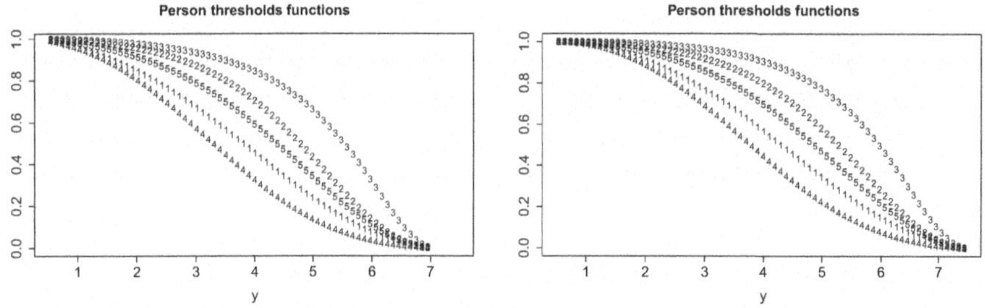

Fig. 6.8 PT functions for political fear data with logit difficulty functions, logistic response function, varying slopes, fixed discrimination parameters, $\theta_p = 0$, left: discrete distribution, right: continuous distribution

functions are very similar. With regard to the fitted functions the approximation by a continuous model seems acceptable. □

In Example 6.2 7-point Likert scales have been used and the logit difficulty function for continuous models used the interval $(1 - c, 7 + c)$, where $c = 0.5$, to obtain $(0.5, 7.5)$. In the discrete case the interval is chosen by $(1 - c, 7.01)$. To illustrate the effect of c in Fig. 6.9 AIC values for varying values c are given (logit difficulty function, normal response function, $\alpha_i = 1$). It is seen that the AIC for the discrete model (drawn line, diamonds) is very stable for varying values of c. The AIC is almost the same as that of corresponding graded response model (3477.146). If a continuous model is fitted (dashed line, circles) for smaller values of c the AIC is approximately the same as for the discrete model. However, AIC increases with increasing values c. For very large values of c the model becomes an approximation of the linear model since the middle part of the logit difficulty function is approximately linear. Thus, for larger values of c the fit deteriorates and the AIC approaches the much worse value of the inappropriate linear model, which is included as a horizontal line.

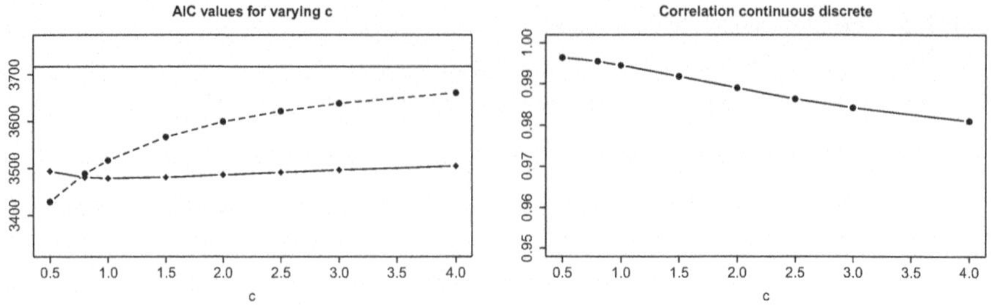

Fig. 6.9 Effect of approximation for fears data, drawn line with diamonds is discrete model, dashed line with circles is continuous model, straight line is linear model

6.6 Flexible Difficulty Functions: A Basis Functions Approach*

In the previous sections difficulty functions were assumed to have the form $\delta_i(y) = \delta_{0i} + \delta_i g(y)$ with fixed function $g(.)$. The model class is already very large and accounts for various distributions, in particular distributions with a restricted support. A strength of the approach is that each item is determined by just two parameters δ_{0i}, δ_i. A potential disadvantage is that, even if one fits models with differing functions $g(.)$, the true distributions of the responses might be not approximated well enough.

A more flexible approach that avoids that one has to choose a specific type of function, and lets the data themselves decide is obtained by letting difficulty functions be determined by basis functions, an approach that has been extensively used in statistics and machine learning (Wood, 2006a, b; Ruppert et al., 2009; Wand, 2000; Hastie and Tibshirani, 1986). Let us assume that the difficulty functions are given by

$$\delta_i(y) = \sum_{l=0}^{M} \delta_{il} \Phi_{il}(y), \tag{6.16}$$

where $\Phi_{il}(.), l = 0, \ldots, M$ are chosen basis functions. The simple choice $\Phi_{i0}(y) = 1$, $\Phi_{i1}(y) = y$, $M = 1$, means that the item difficulty functions are linear. Much more flexible models are obtained by alternative functions as radial basis functions or spline functions. A particular attractive choice are B-splines as propagated and motivated extensively by Eilers and Marx (1996, 2021). They are very flexible and can closely approximate a variety of functions. Since they are very useful they are briefly described in the following.

B-Splines

B-splines are piecewise polynomials and are characterized by

the degree d of the polynomials that are used or, alternatively the order, $o = d + 1$,

a set of knots $\tau_i, \ldots \tau_{i+k+1}$, which are defined on the range of the responses.

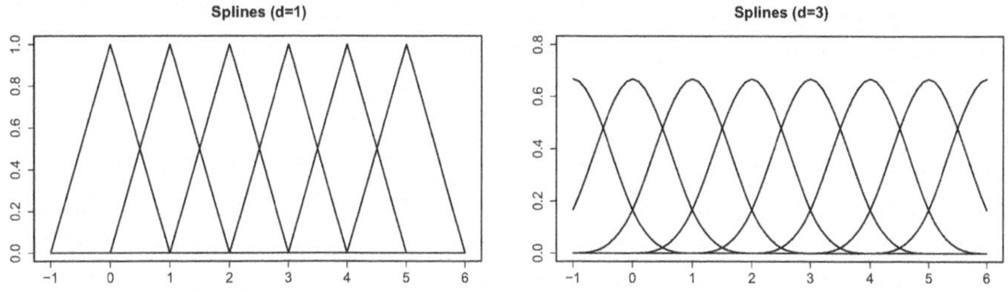

Fig. 6.10 B-splines of degrees 1 (left) and 3 (right) for equally spaced knots

For illustration Fig. 6.10 shows linear splines ($d = 1, o = 2$, left) and cubic splines ($d = 3, o = 4$, right) for inner knots $0, \dots, 5$, which are from the assumed range of responses $[0, 5]$. The other knots (-1 and 6 for linear splines) are additional outer knots, which are not in the range of responses, $[0, 5]$.

The essential properties of B-splines of degree d (order $d + 1$) may be summarized as follows:

- A B-spline consists of $d + 1$ polynomial pieces, each of degree k.
- The polynomial pieces join at d inner knots; at the joining knots, derivates up to order $d - 1$ are continuous.
- The spline function is positive on a domain spanned by $d + 2$ knots; elsewhere it is zero.
- At a given x within the range of responses $d + 1$ B-splines are non-zero.
- Except at boundaries, the spline function overlaps with $2d$ polynomial pieces of its neighbors.
- B-splines sum up to 1 at each point, $\sum_i B_i(x) = 1$.

More concise, a *B-spline function of degree d* with inner knots $\tau_i, \dots \tau_{i+d+1}$ from the domain $[a, b]$ is defined recursively by

$$B_{i,1}(x) = \begin{cases} 1 \text{ if } x \ \in [\tau_i, \tau_{i+1}] \\ 0 \text{ otherwise,} \end{cases}$$

$$B_{i,r+1}(x) = \frac{x - \tau_i}{\tau_{i+r} - \tau_i} B_{i,r}(x) + \frac{\tau_{i+r+1} - x}{\tau_{i+r+1} - \tau_{i+1}} B_{i+1,r}(x),$$

$r = 1, \dots, d$ (Dierckx (1993), p. 8). The last function of this iteration, $B_{i,d+1}(x)$, represents the B-spline function of degree d (order $d + 1$), where additional knots $\tau_{-d} \leq \dots \leq \tau_0 = a, b = \tau_{m_s+1} \leq \dots \leq \tau_{m_s+d+1}$ are used.

Difficulty Functions by Basis Expansions

Basis functions have typically been used to approximate functions of observable variables but here they are used to specify the unobservable difficulty functions. If difficulty functions are expanded in basis functions they have to be non-decreasing, which typically calls for some restrictions. In the case of B-splines a restriction that ensures that functions are non-decreasing is that $\delta_{i0} \leq \cdots \leq \delta_{iM}$.

If a basis has been chosen there are two basic strategies to select the number of basis functions. One is to choose a relatively small number of basis functions, say 6 or 8, which often is enough to provide the needed flexibility. An alternative strategy is to choose a relatively large number, say 30 to 40 basis functions. Then the number of coefficients to be estimated increases strongly and simple log-likelihood fitting is no longer appropriate since it typically yields overfitting. Instead of fitting by maximizing the log-likelihood one has to use penalization methods, that is, one maximizes a penalized log-likelihood, in which the differences between coefficients of adjacent basis functions are restricted to not vary too strongly, see Eilers and Marx (1996, 2021). Disadvantages are that one has to choose a tuning parameter, for example, by cross validation, and that one has to deal with a penalized log-likelihood instead of the usual likelihood. Therefore, we use the former strategy in the examples. More general regularization methods that could be adapted to the smoothing of difficulty functions have been propagated by Eilers and Marx (1996); Hastie and Tibshirani (1986), Bühlmann and Van De Geer (2011). Manuguerra et al. (2020) used monotone I-splines when modeling visual analogue scales.

Although computation is more demanding, flexible difficulties can be used as a diagnostic tool to investigate if a fixed difficulty function is appropriate. It can also be used to investigate if single items have quite different distributions. One should distinguish between two cases, difficulty function as specified in Eq. (6.16), which vary freely across items, and a slightly more restrictive approach, which assumes that only the location varies across items. The latter uses the simpler expansion

$$\delta_i(y) = \delta_{0i} + \sum_{l=0}^{M} \delta_l \Phi_l(y). \tag{6.17}$$

It assumes that the location δ_{0i} is item-specific, but the form of the function is the same for all items.

If difficulty functions have the form (6.16) the model is parametrized by $\alpha_i, \delta_{i0}, \ldots, \delta_{iM}, i = 1, \ldots, I$, as item parameters, and θ_p, $p = 1, \ldots, P$, as person parameters. If they have the form (6.17) the item parameters are $\alpha_i, \delta_{i0}, i = 1, \ldots, I$, $\delta_l, l = 0, \ldots, M$.

The basis function approach also includes the case of fixed difficulty functions. The choice $\Phi_{i0}(y) = 1$, $\Phi_{i1}(y) = g(y)$, $M = 1$, yields the thresholds model TM(F, g). That means the functions 1 and $g(.)$ are considered basis functions. One advantage of representing the models with fixed basis functions as special cases of the basis function approach is that it allows for a common framework of estimation that uses basis functions.

Example 6.3 Fears data (Tutz, 2022b)
Fig. 6.11 shows the PT functions and the fitted difficulty functions for the fear data if B-splines (6 cubic spline functions) generate the difficulty functions, and a discrete distribution is used. For comparison the second row shows the PT and difficulty

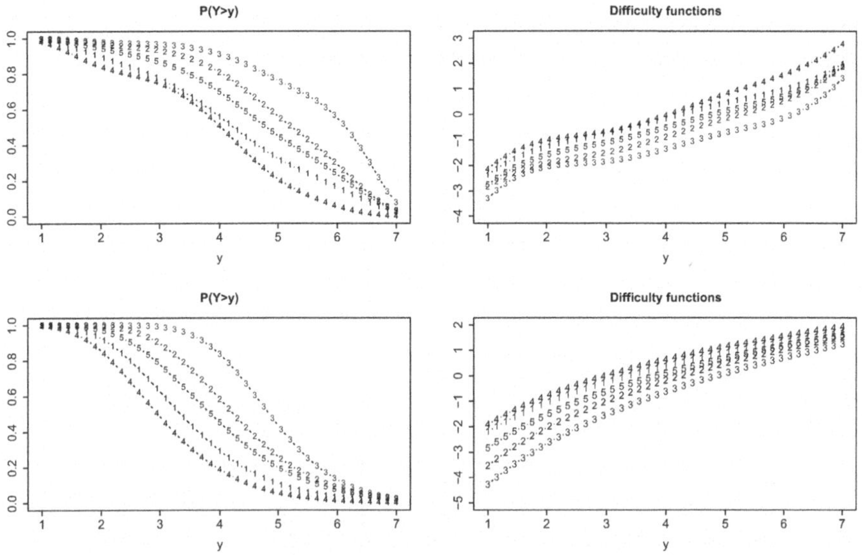

Fig. 6.11 Fitted PT (left) and difficulty (right) functions with B-spline based difficulty functions for fear data; for comparison the second row shows the difficulty functions for logarithmic difficulty functions. Reprinted with permission from Tutz (2022b)

functions for logarithmic difficulty functions. Though the order of the items remains the same, the form of the difficulty functions changes if splines are used instead of the logarithmic function. Also the AIC (3484.07) is distinctly smaller than the value obtained for logarithmic difficulty functions (3980.79). The correlation between posterior estimates of person parameters obtained for the splines and the logarithmic model was 0.959. □

6.7 Choice of Response and Difficulty Function

The thresholds model is mainly determined by the response function $F(.)$ and the item difficulty functions $\delta_i(.)$. In particular the difficulty function determines crucially the range of the response and therefore should be adapted to the item type. Table 6.3 gives an overview of possible combinations of item types and difficulty functions, it is assumed that the response function $F(.)$ is continuous with support $\mathbb{R}$.

If there is reason to assume that responses follow a specific continuous distribution function $F_{\text{resp}}(.)$, for example the normal distribution, the thresholds model with response function $F(y) = 1 - F_{\text{resp}}(-y)$ and linear difficulty functions is a natural choice. Then the distribution of Y_{pi} has distribution function $F_{\text{resp}}(.)$ with the mean and the variance determined in a simple way (Eqs. (6.6) and (6.7)). If $F_{\text{resp}}(.)$ is sym-

Table 6.3 Item types and difficulty functions

Item type	Support	Difficulty function, $\delta_i(y) = \delta_{0i} + \delta_i g(y)$	
Continuous	$Y_{pi} \in \mathbb{R}$	Linear	$g(y) = y$
	$Y_{pi} \geq 0$	Logarithmic	$g(y) = \log(y)$
	$Y_{pi} \in (a, b)$	Logit	$g(y) = \log((y - a)/(b - y)))$
		Inverse	$g(y) = F_{\text{diff}}^{-1}((y - a)/(b - a))$
	$Y_{pi} \in [a, b]$	Logit	$g(y) =$ $\log((y - a + c)/(b + c - y))), c > 0$
		Inverse	$g(y) =$ $F_{\text{diff}}^{-1}((y - a + c)/(b - a + 2c)), c > 0$
Discrete	$Y_{pi} \in \{0, 1\}$		$g(y)$ not used
	$Y_{pi} \in \{0, 1, \dots\}$	Logarithmic	$g(y) = \log(1 + y)$
	$Y_{pi} \in \{0, 1, \dots, k\}$	Logit	$g(y) = \log((y + c)/(k_1 - y))$, $k_1 = k + c, c > 0$
			$g(y) = F_{\text{diff}}^{-1}((y + c)/k_1)$

metric $F(.)$ and $F_{\text{resp}}(.)$ are identical. Then one simply chooses $F_{\text{resp}}(.)$ as response function.

For items with restricted responses the difficulty has to be chosen accordingly. If, for example, $Y_{pi} > 0$ then difficulty functions should be chosen such that $\lim_{y \to 0} \delta_i(y) = -\infty$ to ensure that $Y_{pi} > 0$ holds. This is the case for the logarithmic function included in Table 6.3. If responses are from a known interval (a, b) difficulty functions should fulfill $\lim_{y \to a} \delta_i(y) = -\infty$, $\lim_{y \to b} \delta_i(y) = \infty$. Natural choices are inverse distribution functions, $g(y) = F_{\text{diff}}^{-1}((y - a)/(b - a))$, which in the special case of the logistic distribution function becomes $g(y) = \log((y - a)/(b - y)))$. After a transformation of the responses into the interval $(0, 1)$ one obtains the logit transformation $g(y) = \log(y/(1 - y))$. If the boundaries are included, that is $Y_{pi} = a$ and $Y_{pi} = b$ are observed the difficulty functions are better defined on the wider interval $(a - c, b + c)$, where c is a small constant. The corresponding difficulty functions are also given in the table.

If responses are discrete and have finite support the response distribution is the multinomial distribution. If the response is binary, the choice of the difficulty function is without consequences, any increasing function can be used since the model contains only one item parameter given by $\delta_i(0)$. If $Y_{pi} \geq 0$, as in count data, the restriction is fulfilled for the (transformed) logarithmic function $g(y) = \log(1 + y)$. If $Y_{pi} \in \{0, 1, \dots, k\}$, an adapted inverse (adapted) distribution function can be used, for example the logit function. In Table 6.3 the interval $(1 - c, k_1)$ is used to define the corresponding difficulty function, where k_1 is slightly greater than k, for example, $k_1 = k + 0.01$ such that the function is properly defined.

6.8 Combining Different Types of Items

A strength of the thresholds model is that different item formats can be used within one test and be linked to the same latent trait. Tests can contain binary items, polytomous items, continuous ones as well as count data items. The differences in item formats are captured in the difficulty functions. They determine which responses can be expected given a fixed ability parameter, and what distributional form the responses have. By construction the model assumes that there is a common latent trait that determines the outcome for all items. The model has the same form for continuous, polytomous or binary items, only the difficulty functions have to be adapted to the items. There is no common difficulty function that holds for all items if they have different formats. Thus, when items with different formats are in a test one chooses difficulty functions as given in Table 6.3 according to the specific items.

There is no need to refer to linkage methods for mixed-format tests as proposed, for example by Kim and Lee (2006), where linkage methods for data obtained from two groups of examinees through common items were proposed. Common linkage methods are mean/mean, mean/sigma, and Stocking-Lord linkage, see, for example, (Hanson and Béguin, 2002; Ogasawara, 2001; Kim and Hanson, 2002; Kolen and Brennan, 2014). The thresholds model allows to use the same person parameters in differently scaled items, therefore linkage methods are unnecessary.

The following example illustrates the fitting with mixed format items. In addition to a set of ordered items with three categories a continuous item is included. The advantage of including continuous response items is that they typically contain more information than binary items or items with a small number of response categories.

Example 6.4 Verbal aggression
As an illustrating example we consider the verbal aggression data VerbAgg from package *lm4*, which have been used by De Boeck and Wilson (2004) to illustrate various forms of item response models. We follow their description of the data set. The data are from 316 respondents presented with a verbal aggression inventory with 24 items. Each item consists of three verbally aggressive behaviors (cursing, scolding, wanting), distinguishes between self-to-blame and others-to blame, and two behavioral modes (wanting, doing). A sample item that refers to cursing, other-to-blame and wanting is "A bus fails to stop for me. I would want to curse". We use the six items that refer to behavior mode wanting and others-to blame which are ordered factors with levels no, perhaps, yes. In addition to the categorical items, the subject's trait anger score as measured on the State-Trait Anger Expression Inventory (STAXI) is given. The latter is considered the first item and the six categorical items as items 2–7.

For the first item the logarithmic difficulty function has been used since aggression scores are positive and continuous (values between 11 and 39). For the categorical items logit difficulty functions have been used to ensure that the response is between 0 and 2. As response function the logistic distribution function has been used.

Table 6.4 Estimated parameters for aggression data

Item	Intercept	Slope
1	−14.04	4.732
2	0.309	0.670
3	0.594	0.597
4	0.966	0.620
5	0.171	0.759
6	0.565	0.628
7	0.896	0.568

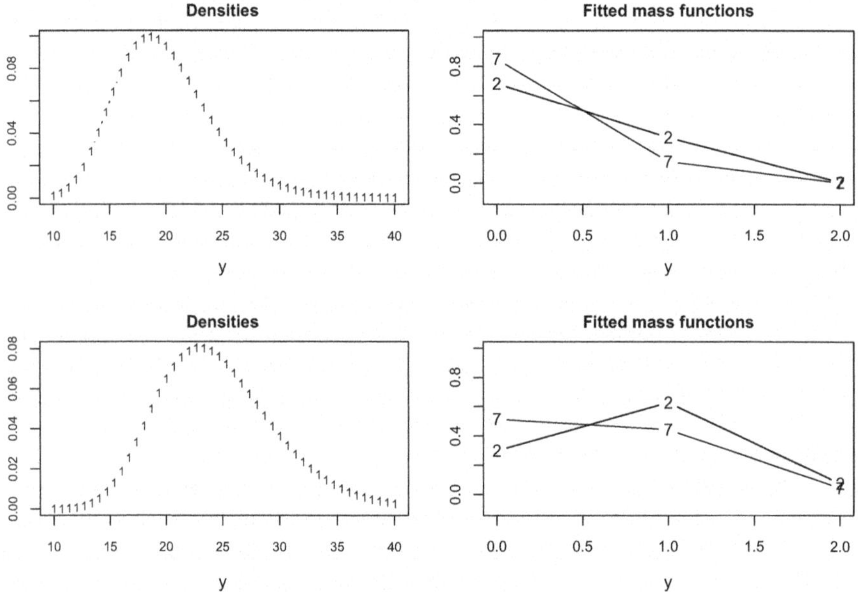

Fig. 6.12 Estimated densities and mass functions for aggression data, left: first item (continuous), right: two categorical items, first row: $\theta = 0$, second row: $\theta = 1$

Table 6.4 shows the estimated parameters of the model, the estimated standard deviation of the mixing distribution was 0.697. It is seen that the estimated parameters for the first item are quite different from the estimates obtained for the categorical items. Figure 6.12 shows the densities (first item) and the mass functions for two categorical items (items 2 and 7). The first row shows the fitted functions if $\theta = 0$, in the second row $\theta = 1$. □

6.9 Estimation and Inference

In the following marginal maximum likelihood methods for the estimation of item parameters and posterior estimation of person parameters are considered under the usual assumption of conditional independence of observable variables given the latent variables.

Marginal Maximum Likelihood Estimation
Let the general thresholds model hold. Then the distribution function for observation Y_{pi} has the form

$$F_{pi}(y) = P(Y_{pi} \le y) = 1 - F(\alpha_i(\theta_p - \delta_i(y))).$$

For *continuous* responses, one obtains the density by building derivatives yielding

$$f_{pi}(y) = \frac{\partial F_{pi}(y)}{\partial y} = f(\alpha_i(\theta_p - \delta_i(y)))\alpha_i \delta_i'(y),$$

where $f(.)$ is the density corresponding to $F(.)$, and $\delta_i'(y) = \partial\delta_i(y)/\partial y$ is the derivative of the threshold function.

For *discrete* responses $Y_{pi} \in \{0, 1, \dots\}$ the probability mass function is obtained by building differences. Then, one has the discrete density function

$$f_{pi}(0) = 1 - P(Y_{pi} > 0) = 1 - F(\alpha_i(\theta_p - \delta_i(0))),$$
$$f_{pi}(r) = P(Y_{pi} > r - 1) - P(Y_{pi} > r) = F(\alpha_i(\theta_p - \delta_i(r - 1))) - F(\alpha_i(\theta_p - \delta_i(r))), r = 1, 2, \dots$$

where $\sum_r f_{pi}(r) = 1$. For simple binary responses one obtains

$$f_{pi}(0) = 1 - F(\alpha_i(\theta_p - \delta_i(0))) \quad f_{pi}(1) = F(\alpha_i(\theta_p - \delta_i(0))),$$

where $\delta_{0i} = \delta_i(0)$ is the familiar difficulty parameter.

If difficulties are expanded in basis functions they have the form

$$\delta_i(y) = \sum_{l=0}^{M} \delta_{il}\Phi_{il}(y) = \boldsymbol{\Phi}_i(y)^T \boldsymbol{\delta}_i,$$

where $\boldsymbol{\Phi}_i(y)^T = (\Phi_{i0}(y), \dots, \Phi_{iM}(y))$, $\boldsymbol{\delta}_i^T = (\delta_{i0}, \dots, \delta_{iM})$. The corresponding derivative is given by

$$\delta_i'(y) = \sum_{l=0}^{M} \delta_{il}\Phi_{il}'(y) = \boldsymbol{\Phi}_i'(y)^T \boldsymbol{\delta}_i,$$

where $\boldsymbol{\Phi}_i'(y)^T = (\Phi_{i0}'(y), \dots, \Phi_{iM}'(y))$ is the vector of derivatives of basis functions.

Let now observations be given by $y_{pi}, i = 1, \dots, I, p = 1, \dots, P$. The estimation method that is used is marginal likelihood by assuming that person parameters are normally distributed, $\theta_p \sim N(0, \sigma_\theta^2)$. Maximization of the marginal log-likelihood can be obtained by integration techniques. We use numerical integration by Gauss-Hermite integration methods. Early versions for univariate random effects date back to Hinde (1982) and Anderson and Aitkin (1985).

Let $\boldsymbol{\delta}_i$ denote the vector of all parameters linked to the difficulty function of item i. For fixed difficulty functions it has length two, for expansions in basis functions it is, more generally, $M + 1$. The vector $\boldsymbol{\alpha}^T = (\alpha_1, \dots, \alpha_I)$ collects the item slopes. With $\boldsymbol{\delta}^T = (\boldsymbol{\delta}_1^T, \dots, \boldsymbol{\delta}_I^T, \boldsymbol{\alpha}^T, \sigma_\theta)$ denoting the set of all item parameters and $f_{0,\sigma_\theta}(.)$ denoting the density of the normal distribution $N(0, \sigma_\theta^2)$ the marginal likelihood has the form

$$L(\boldsymbol{\delta}) = \prod_{p=1}^{P} \int \prod_{i=1}^{I} f_{pi}(y_{pi}) f_{0,\sigma_\theta}(\theta_p) d\theta_p,$$

yielding the log-likelihood

$$l(\boldsymbol{\delta}) = \log(L(\boldsymbol{\delta})) = \sum_{p=1}^{P} \log\left(\int \prod_{i=1}^{I} f_{pi}(y_{pi}) f_{0,\sigma_\theta}(\theta_p) d\theta_p\right).$$

The score function $s(\boldsymbol{\delta}) = \partial l / \partial \boldsymbol{\delta}$, which takes on the value of 0 for maximum likelihood estimates, has components

$$\frac{\partial l}{\partial \delta_{ij}} = \sum_{p=1}^{P} \int \frac{\partial f_{pi}(y_{pi})}{\partial \delta_{ij}} \prod_{l \neq i} f_{pl}(y_{pl}) f_{0,\sigma_\theta}(\theta_p) d\theta_p / c_p,$$

$$\frac{\partial l}{\partial \alpha_i} = \sum_{p=1}^{P} \int \frac{\partial f_{pi}(y_{pi})}{\partial \alpha_i} \prod_{l \neq i} f_{pl}(y_{pl}) f_{0,\sigma_\theta}(\theta_p) d\theta_p / c_p,$$

$$\frac{\partial l}{\partial \sigma_\theta} = \sum_{p=1}^{P} \int \prod_{i=1}^{I} f_{pi}(y_{pi}) \frac{\partial f_{0,\sigma_\theta}(\theta_p)}{\partial \sigma_\theta} d\theta_p / c_p,$$

where $c_p = \int \prod_{i=1}^{I} f_{pi}(y_{pi}) f_{0,\sigma_\theta}(\theta_p) d\theta_p$. The derivation uses that the order of integration and differentiation can be interchanged, which holds if densities are continuous and continuously differentiable, in particular it holds if $F(.)$ is the normal distribution function.

The form of the derivatives depends on the distribution of the responses. For continuous responses one obtains

$$\frac{\partial f_{pi}(y_{pi})}{\partial \delta_{ij}} = \alpha_i \Phi'_{ij}(y_{pi}) f(\alpha_i(\theta_p - \delta_i(y_{pi}))) - \alpha_i^2 f'(\alpha_i(\theta_p - \delta_i(y_{pi}))) \Phi_{ij}(y_{pi}) \boldsymbol{\Phi}'_i(y_{pi})^T \boldsymbol{\delta}_i,$$

$$\frac{\partial f_{pi}(y_{pi})}{\partial \alpha_i} = \boldsymbol{\Phi}'_i(y_{pi})^T \boldsymbol{\delta}_i \{ f(\alpha_i(\theta_p - \delta_i(y_{pi})) + \alpha_i(\theta_p - \boldsymbol{\Phi}_i(y_{pi})^T \boldsymbol{\delta}_i) f'(\alpha_i(\theta_p - \delta_i(y_{pi}))) \},$$

with $f'(.)$ denoting the derivative of $f(.)$. For discrete responses one has

$$\frac{\partial f_{pi}(y_{pi})}{\partial \delta_{ij}} = -\alpha_i f(\alpha_i(\theta_p - \delta_i(y_{pi}-1)))\Phi_{ij}(y_{pi}-1) + \alpha_i f(\alpha_i(\theta_p - \delta_i(y_{pi})))\Phi_{ij}(y_{pi}),$$

$$\frac{\partial f_{pi}(y_{pi})}{\partial \alpha_i} = (\theta_p - \delta_i(y_{pi}-1)) f(\alpha_i(\theta_p - \delta_i(y_{pi}-1))) - (\theta_p - \delta_i(y_{pi})) f(\alpha_i(\theta_p - \delta_i(y_{pi}))),$$

where $\Phi_{ij}(-1)$ is defined by $\Phi_{ij}(-1) = 0$ and $\delta_i(-1) = -\infty$.

For simple difficulty functions the score functions simplify accordingly. For example, when the difficulty functions are linear one has $\boldsymbol{\Phi}_i(y)^T = (1, y)$, and $\boldsymbol{\Phi}_i'(y)^T = (0, 1)$. An approximation of the covariance of the estimate, $\text{cov}(\hat{\boldsymbol{\delta}})$, is obtained by the observed information $-\partial^2 l/\partial \boldsymbol{\delta} \partial \boldsymbol{\delta}^T$.

Some caution is needed when fitting the model (6.17) with a common difficulty function expanded in B-splines. Since B-splines sum up to 1 at any given value the parameters in model (6.17) are not identified. This can be fixed by choosing a fixed value for one of the parameters δ_{0i}, for example, $\delta_{01} = 0$. One can also use the general form (6.16) and use a tailored penalty. Instead of maximizing the log-likelihood one maximizes the penalized log-likelihood $l(\{\delta_i\}) = l(\{\delta_i\}) - P_\lambda(\{\delta_i\})$ with penalty term

$$P_\lambda(\{\delta_i\}) = \lambda \sum_{i=2}^{I} \sum_{l=2}^{M} [(\delta_{il} - \delta_{i,l-1}) - (\delta_{i-1,l} - \delta_{i-1,l-1})]^2.$$

The choice of λ determines the estimates. For $\lambda = 0$ one maximizes the usual log-likelihood. For $\lambda \longrightarrow \infty$ the differences of adjacent parameters become the same for all items, that is, $\delta_{il} - \delta_{i,l-1} = \delta_{i-1,l} - \delta_{i-1,l-1}$, otherwise the penalty term would also go to infinity. Then, the levels of the functions can differ but not the form of the function. Therefore, using a large value of λ automatically yields shifted difficulty functions.

For the computation of estimates we used Gauss-Hermite integration, which works rather well since all integrals are unidimensional. The written R program uses the computation of derivatives in combination with the R function *optim*.

Estimating Person Parameters

If estimates of item parameters are found, posterior mode or mean estimation yields estimates of person parameters. For given item responses $\boldsymbol{y}_p^T = (y_{p1}, \ldots, y_{pI})$, the posterior is given by

$$f(\theta_p | \boldsymbol{y}_p, \boldsymbol{\delta}) = \frac{\prod_{i=1}^{I} f(\alpha_i(\theta_p - \delta_i(y_{pi})))\alpha_i \delta_i'(y_{pi}) f_{0,\sigma_\theta}(\theta_p)}{\int \prod_{i=1}^{I} f(\alpha_i(\theta_p - \delta_i(y_{pi})))\alpha_i \delta_i'(y_{pi}) f_{0,\sigma_\theta}(\theta_p) d\theta_p}.$$

Replacing the parameter $\boldsymbol{\delta}$ by its estimate $\hat{\boldsymbol{\delta}}$ allows to compute the mode of $f(\theta_p|\boldsymbol{y}, \boldsymbol{\delta})$ or the posterior mean

$$\hat{\theta}_p^{(m)} = \mathrm{E}(\theta_p|\boldsymbol{y}_p, \boldsymbol{\delta}) = \int \theta f(\theta|\boldsymbol{y}_p, \boldsymbol{\delta}) d\theta.$$

Fitting Thresholds Models
For the fitting of thresholds model special software is needed. The fitting of some of the examples considered in this chapter can be obtained by using software provided on the website for this book.

6.10 Measuring Physical Fitness

Most examples in this book refer to the measurement of mental abilities or attitudes. In the final example of this chapter a quite different latent trait is considered: physical fitness. It is interesting since physical abilities as fitness have hardly been modeled as a latent trait. Performance analysis of sports and laboratory data typically focuses on the observations themselves and examines physical performance tests as ergometer data measured repeatedly on an individual (Hopkins et al., 2001). Sometimes test batteries have been used but typically simply the sum of test item scores is used. Scores can be distances when throwing a ball and time to run a specific distance. To obtain a sum score test items that are measured as the time needed to accomplish the items are simply converted to 1/score (Fjørtoft et al., 2011). Building scores in this way without any weights or transformation of item scores just assumes that it measures the right thing but there is no measurement model involved that links a trait to observations. A possible reason why latent trait models have not been used to measure physical abilities is that observations typically are continuous and no efficient latent trait model has been available. However, the thresholds model explicitly can handle continuous responses. In the following we briefly consider the measurement of physical abilities by using decathlon data, for more details, see Tutz (2024b).

Example 6.5 Decathlon data
We use the decathlon data (data set decathlon) which are included in the R package FactoMineR. When modeling decathlon data one can hardly expect the typically used normal distribution to work well. The reason is that in physical performance tests there are often thresholds that can hardly be overcome. For example, in jumping there typically is an upper limit that hardly any athlete can overcome but lower values can and do occur. Consequently, in jumping data one can expect left skewed distributions but hardly a normal distribution.

In a decathlon three jumping tasks are included: long jump, high jump, and pole vault. We fitted several models, one with a normal distribution as response function, one that uses the right-skewed Gumbel as the distribution of the responses,

Table 6.5 Fitting of jumping data. Reprinted with permission from Tutz (2024b)

Response distribution	Log-likelihood	Std mixture
Normal	26.628	0.362
Gumbel	24.736	0.406
Gompertz	18.813	0.652

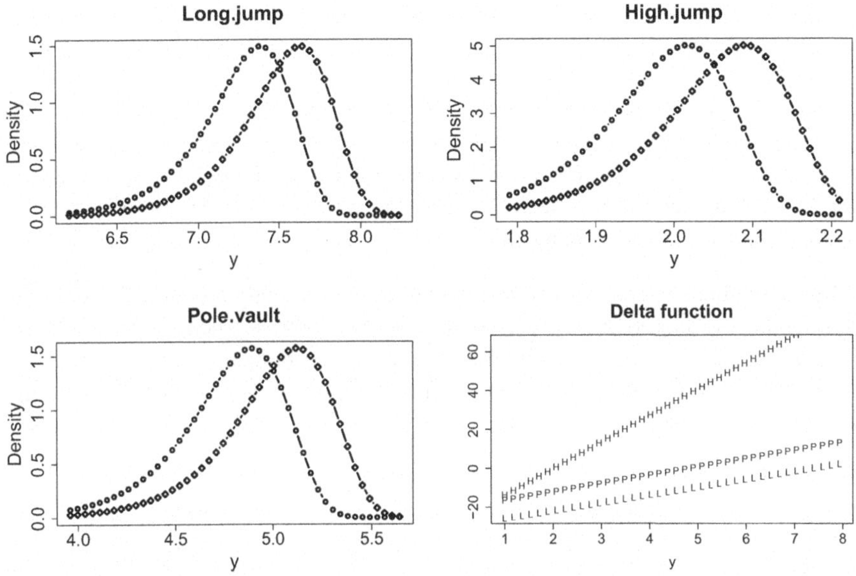

Fig. 6.13 Response distributions for jumping data (Gompertz model) for $\theta_p = 0$ (dots) and $\theta_p = 1$ (diamonds). Reprinted with permission from Tutz (2024b)

and one that uses the left-skewed Gompertz distribution (linear difficulty functions, $\alpha_i = 1$). Table 6.5 shows the resulting log-likelihoods and the standard deviation of the mixing distribution when marginal estimation is used. It is seen that the left-skewed Gompertz distribution yields a much better fit than the other distributions. Figure 6.13 shows the resulting distributions for the Gompertz model for $\theta_p = 0$ (dots) and $\theta_p = 1$ (diamonds). It is seen that a person with larger ability has much better chances of jumping wider and higher. The distributions are distinctly not normally distributed.

Figure 6.13 also shows the fitted difficulty functions for the three tasks (H for high jump, L for long jump, and P for pole vault). It is seen that the lines are almost parallel for log jump and pole vault, slopes are almost the same (4.05 and 4.26) indicating that the variances are almost the same. The slope for high jump is much larger (13.63) indicating that the variance is smaller, which is to be expected since the range of values for high jump is much smaller than for long jump and pole vault. It should also be noted that, although the difficulty functions are plotted for the range

$(1, 8)$, the range that is relevant in the task high jump is much smaller, between 1.8 and 2.2. It is a strength of the model that it automatically accounts for the differing ranges of values in different tasks.

Modeling running data is different from modeling jumping data in several respects. Running 100 m in less than 10 s is hardly possible while there is no upper limit to time involved. Therefore, such data can be expected to be right skewed in contrast to jumping data where there is an upper limit. In addition the parameter θ_p has a different meaning. Large values of θ_p mean that the probability of a response above thresholds is large but that means poor, not good performance. Therefore, it is an inability parameter, however one that is easily transformed into an ability parameter by considering $-\theta_p$. Then, large values again mean a tendency to good performance. For running data the right-skewed Gumbel distribution shows the best fit. Figure 6.14 shows the resulting distributions for the model for $\theta_p = 0$ (dots) and $\theta_p = 1$ (diamonds). It is seen how running time increases if ability is smaller (inability larger, $\theta_p = 1$) than average. □

In classical latent trait models the range of items is usually the same for all tasks, either binary or categorical with a fixed number of categories. Then, it makes sense to compare and interpret the difficulty functions as representing the varying difficulties of the tasks. When investigating, for example, jumping ability the tasks wide jump, high jump, and pole vault are by nature different. It is hardly sensible to compare

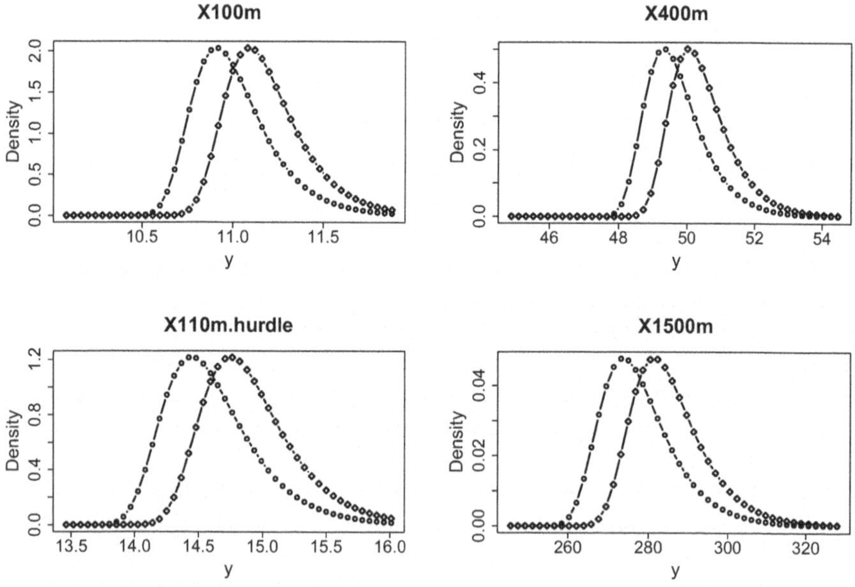

Fig. 6.14 Response distributions for running data (Gumbel model) for $\theta_p = 0$ (dots) and $\theta_p = 1$ (diamonds). Reprinted with permission from Tutz (2024b)

them in terms of difficulty, high jump is not harder than wide jump. It depends on the concrete values which task is harder, which is considered briefly in the following.

The difficulty functions in the thresholds model provide a tool to compare the difficulties of specific tasks for concrete values. Let us consider task i and j and consider the pair of values $(y^{(i)}, y^{(j)})$ for which

$$P(Y_{pi} > y^{(i)}|\theta_p, \delta_i) = P(Y_{pj} > y^{(j)}|\theta_p, \delta_j).$$

It means the probability of a performance above threshold $y^{(i)}$ in task i is the same as the probability of a performance above threshold $y^{(j)}$ in task j. It is easily derived that the tuples are given by

$$IP_{ij} = \{(y^{(i)}, y^{(j)}) : y^{(j)} = (\delta_{0i} + \delta_i y^{(i)} - \delta_{0j})/\delta_j\}.$$

which is called the *isoprobability curve of tasks i and j*. Since difficulty functions are linear, the curves are straight lines. The curves do not depend on the ability θ_p. For smaller θ_p the probabilities $P(Y_{pi} > y^{(i)}|\theta_p, \delta_i) = P(Y_{pj} > y^{(j)}|\theta_p, \delta_j)$ will be smaller than for large θ_p but the curves do not depend on θ_p.

In Fig. 6.15, the isodifficulties are given for the jumping data. It is seen how the probability of jumping wider than a specific value is linked to jumping higher than specific values. For example, for any θ_p the probability of jumping wider than 7.32 m equals the probability of jumping higher than 2 m.

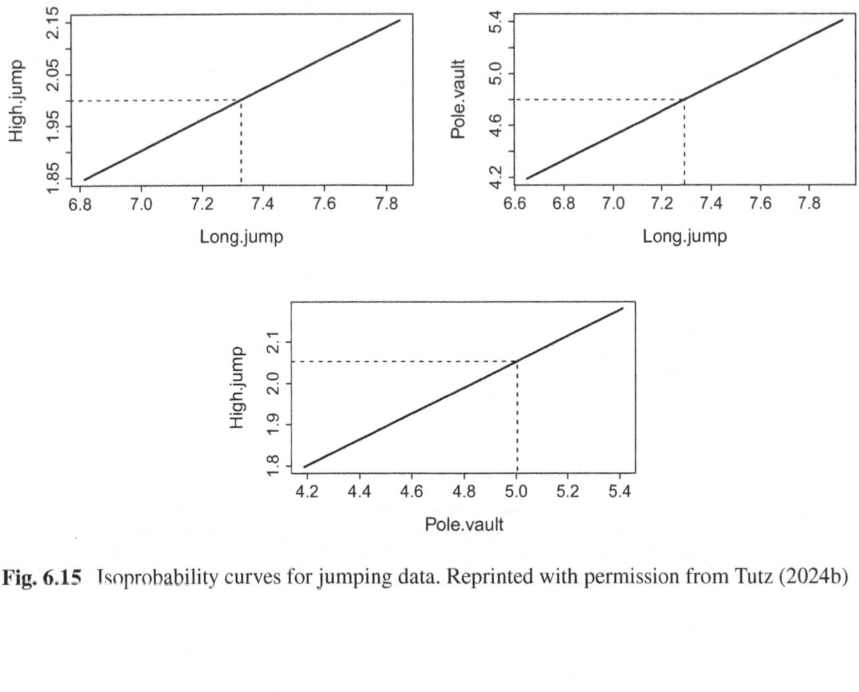

Fig. 6.15 Isoprobability curves for jumping data. Reprinted with permission from Tutz (2024b)

The isoprobability curve allows to compare the difficulties of different task. In particular the slope of the linear function that defines the isodifficulties is interesting. For the isodensities (long jump, high jump) the slope is 0.30, that means an increase of the performance by one unit of the task long jump is equivalent to an increase of the performance 0.30 in units of the performance in the task high jump. For example, an increase of 0.5 m in long jump corresponds to an increase of 0.15 m in the task high jump.

Chapter 7
Classical Test Theory

Classical test theory (CTT) was the dominating theory of measurement before the development of latent trait theory and is still used in the measurement of proficiency and assessments of attitude. The main feature of CTT is that it postulates the existence of a true score, that error scores are uncorrelated with each other and with true scores and that observed, true and error scores are linearly related (Novick, 1966; Lord and Novick, 1968).

In the following we consider classical test theory only briefly. The model and basic concepts as reliability and validity are given and basic differences between CCT and modern latent trait theory are highlighted. We also consider genuine latent trait models that can be seen as background models for the CCT model.

7.1 The Classical Test Theory Model

Let X_{pm} denote the score of a person p on measurement m. With τ_{pm} denoting the expected value of the observed score, $\tau_{pm} = \mathrm{E}(X_{pm})$, the score can be decomposed into two components,

$$X_{pm} = \tau_{pm} + E_{pm},$$

where E_{pm} is the error-score variable given by $E_{pm} = X_{pm} - \tau_{pm}$. The error score is a random variable with $\mathrm{E}(E_{pm}) = 0$.

Classical test theory addresses individual differences by considering the distribution of measurements over persons. Thus, let X_{*m} denote the score for a randomly selected test taker, T_{*m} the corresponding true score, and E_{*m} the corresponding error. The subscript p is replaced by the subscript $*$ to indicate that a person has been randomly selected from a specified population. One obtains the decomposition

$$X_{*m} = T_{*m} + E_{*m}, \tag{7.1}$$

G. Tutz, *A Short Guide to Item Response Theory Models*, Statistics for Social and Behavioral Sciences, https://doi.org/10.1007/978-3-031-87271-6_7

in which T_{*m} is a random variable and not a fixed value. Under the weak assumption that the variance of the observed scores is finite and X_{pm}, X_{pn} denoting measurements that are assumed to be independent for each p it can be derived that

$$\text{the expected error score is zero, } \mathrm{E}(E_{*m}) = 0, \tag{7.2}$$

$$\text{the correlation between true and error score is zero, } \rho(T_{*m}, E_{*m}) = 0, \tag{7.3}$$

$$\text{the correlation between errors is zero, } \rho(E_{*m}, E_{*n}) = 0, \tag{7.4}$$

$$\text{the correlation between } E_{*m} \text{ and } T_{*n} \text{ is zero, } \rho(E_{*m}, T_{*n}) = 0 \tag{7.5}$$

(see Theorem 2.7.1 Lord and Novick (1968)). The equation $X_{*m} = T_{*m} + E_{*m}$ together with the assumptions (7.2)–(7.5) is considered the *classical test theory model* (Lord and Novick, 1968).

Classical Test Theory model (randomly selected test taker)

$$X_{*m} = T_{*m} + E_{*m},$$

$$\mathrm{E}(E_{*m}) = 0, \quad \rho(T_{*m}, E_{*m}) = 0,$$
$$\rho(E_{*m}, E_{*n}) = 0, \quad \rho(E_{*m}, T_{*n}) = 0$$

Reliability

Let us consider the model for a fixed measurement with randomly selected test takers in the form $X = T + E$, where for simplicity the indices m and $*$ are omitted. The concept of reliability relates to the problem of how well the observed score is able to capture the true score. Formally, *reliability* is the squared correlation between the observed score and the true score $\rho^2_{X,T}$ also known as the coefficient of determination. It is given by

$$\rho^2_{X,T} = \frac{\sigma^2_T}{\sigma^2_X} = 1 - \frac{\sigma^2_E}{\sigma^2_X}, \tag{7.6}$$

where σ^2_X is the variance of X, σ^2_T is the variance of T and σ^2_E is the variance of E, which are linked by $\sigma^2_X = \sigma^2_T + \sigma^2_E$. Reliability is large if the variance ration σ^2_E/σ^2_X is small, that means when the error variance is small as compared to the variance of the observed scores. It is a measure of the strength of the relationship between observed and true scores.

While the variance of the observed score can be estimated it is much harder to estimate the variance of the true score, which is not observed. A way to approximate reliability is to consider so-called *parallel measurements*, that is measurements

$$X = T + E \quad \tilde{X} = T + \tilde{E},$$

with $\mathrm{var}(E) = \mathrm{var}(\tilde{E})$. Measurements X and $\tilde{X}$ can be considered repeated measurements since they measure the same true score T. It can be derived that the reliability is equivalent to the correlation between the parallel measurements $\rho_{X,\tilde{X}}$,

$$\rho^2_{X,T} = \rho_{X,\tilde{X}}.$$

Thus, if parallel measurements are available reliability can be estimated since it is linked to the correlation of two observable variables. The equivalence also allows to draw inferences on the error variance since $\sigma_E^2 = \sigma_X^2(1 - \rho_{X,\tilde{X}})$, which follows from Eq. (7.6).

Validity

While reliability refers to the precision of a test by comparing the error variance and the variance of the observed variable validity typically investigates if the results of an instrument are related to results on other instruments or variables of interest. A classical way to measure the association between two measurements $X, \tilde{X}$ is to use the correlation coefficient

$$\rho_{X,\tilde{X}} = \frac{\mathrm{cov}(X, \tilde{X})}{\sqrt{\mathrm{var}(X)}\sqrt{\mathrm{var}(\tilde{X})}}.$$

The absolute value of $\rho_{X,\tilde{X}}$ is considered the validity coefficient of measurement X with respect to measurement $\tilde{X}$.

The measurement $\tilde{X}$ can refer to some performance in the future, for example, the performance in final exams if X is the performance in entrance tests. Then validity has the form of a *predictive validity*. Another form is *concurrent validity*, which means that X and $\tilde{X}$ are considered alternative ways of measuring the same trait.

If the two measurements are classical test theory measurements with the same underlying true score, $X = T + E, \tilde{X} = T + \tilde{X}$, one obtains $\mathrm{cov}(X, \tilde{X}) = \mathrm{cov}(T, \tilde{T})$, and therefore

$$\rho_{X,\tilde{X}} = \frac{\mathrm{cov}(T, \tilde{T})}{\sqrt{\mathrm{var}(X)}\sqrt{\mathrm{var}(\tilde{X})}}.$$

If $X, \tilde{X}$ are parallel measurements $\mathrm{cov}(T, \tilde{T}) = \mathrm{var}(T)$ and $\mathrm{var}(X) = \mathrm{var}(\tilde{X}) = \sigma_X^2$ hold and one has $\rho_{X,\tilde{X}} = \mathrm{var}(T)/\mathrm{var}(X) = \sigma_T^2/\sigma_X^2$. Thus, the reliability of a test is equivalent to the validity of a test with respect to a parallel test.

Further forms that are less clearly linked to alternative measurements are content validity and construct validity. *Content validity* relies on experts to establish if measurements relate to the trait that is to be measured. *Construct validity* is based on the theoretical understanding of the trait to be measured, it aims at forming a concept of the trait to be measured as a construct. The other forms of validity are sometimes seen as ways to establish construct validity, as considered, for example, by Messick (1989).

7.2 Classical Test Theory and the Measurement Instrument

There are several ways of specifying the classical test theory model. In particular one can start with the sum score of a test or at the item level. In the following both approaches are considered briefly.

Sum Score and Classical Test Theory

Classical test theory typically takes the score of a person on an instrument as the sum of the person's score on the items. Then, in the decomposition $X = T + E$ the observable variable X is the sum score. The model makes rather weak structural assumptions, just a rather trivial decomposition into expectation and error is assumed. However, most often it is assumed that the true and the error score follow a normal distribution.

Apart from the error the observed score is essentially the latent trait and not much can be said about single items since they are not a formal part of the model. In particular there is no item location parameter, which in proficiency tests is interpreted as the difficulty of an item. Thus, the separation of person ability and item difficulty, which is a central part of modern latent trait theory, is missing. The CTT model for sum scores is a simple distribution model that applies to a random sample from the population.

Although item difficulties are not explicitly specified by the model one can try to use the number of persons who answered an item correctly as a measure of easiness if items are dichotomously scored. If items are polytomously scored the average score can be used to measure easiness. The model itself and the computation of easiness scores *implicitly* assume that it makes sense to consider the mean of the sum score, $\mathrm{E}(X)$. Means are certainly sensible if all items are binary items. It is less clear in polytomous items. In proficiency tests scores like $0, 1, 2$ may refer to quite different performance. If, for example, the scores refer to 'not solved/no response at all ', 'partially solved' and 'perfectly solved' and 'partially solved' refers to just a very small step in the process that leads to perfect solution the distance between the performance events can hardly be considered as equal. Then the mean is not an appropriate measure for the performance. The problem is less severe in attitude measurement with Likert scales. If one pragmatically assumes that scores are on an interval scale the use of the mean may be warranted.

Items as Replications

Simple representations of reliability can be obtained if items are seen as replications. If one assumes that each item is a replication of every other item for a randomly selected person one can consider for item i the CCT decomposition

$$X_i = T + E_i \tag{7.7}$$

with the assumptions that $\mathrm{E}(E_i) = 0$, the true score and the error score are uncorrelated and the variances given by $\sigma_E^2 = \mathrm{var}(E_i)$, $\sigma_T^2 = \mathrm{var}(T)$. Then, for the sum score $Y_= \sum_{i=1}^{I} X_i$ the decomposition

$$Y = IT + E$$

holds, where $E = \sum_{i=1}^{I} E_i$. Using $\mathrm{var}(IT) = I^2\sigma_T^2$ and $\mathrm{var}(E) = I\sigma_E^2$ yields the reliability

$$r = \frac{\mathrm{var}(IT)}{\mathrm{var}(X)} = \frac{I^2\sigma_T^2}{I^2\sigma_T^2 + I\sigma_E^2} = \frac{\sigma_T^2}{\sigma_T^2 + \sigma_E^2/I},$$

see also Andrich and Marais (2019). It is seen that as the number of items increases the reliability increases since σ_E^2/I becomes smaller. An estimate of reliability is provided by Cronbach's α, which, following Andrich and Marais (2019), can be given in the form

$$\alpha = \frac{I}{I-1} \frac{s_Y^2 - \sum_{i=1}^{I} s_i^2}{s_Y^2},$$

where $s_Y^2 = \sum_{p=1}^{P}(y_p - \bar{y})$, $s_i^2 = \sum_{p=1}^{P}(x_{pi} - \bar{x})$ are the usual estimates of standard errors with y_p denoting the sum score of person p, and x_{pi} denoting the observed score of person p on item i. This form of reliability is also referred to as internal reliability, which is computed from the items within a test with the test administered just once.

The problem with these derivations is that usually items cannot be considered simple replications. In particular it is assumed that all items have the same difficulty, in the "model" (7.7) the true score does not depend on the item. Also the assumption of the same variance of all error terms is a very strong assumption. If the discrimination of an item is considered the correlation between the scores on the item and the total test score, the model implicitly assumes that all items have the same discrimination. For further problems concerning, for example, independence of items and unidimensionality see Andrich and Marais (2019).

As in the case where the classical model is assumed to hold for the total score it is needed that the mean is a sensible measure for the performance on single items, which is only without problems if items are binary or counts.

Classical Test Theory Model for Single Items

Somewhat weaker one can assume that for single items and randomly selected persons the decomposition

$$X_i = T_i + E_i$$

holds with the assumptions that $\mathrm{E}(E_i) = 0$, and the true score and the error score are uncorrelated. Then, for the sum score $Y = \sum_{i=1}^{I} X_i$ the decomposition

$$Y = T + E$$

holds, where $T = \sum_{i=1}^{I} T_i$ and $E = \sum_{i=1}^{I} E_i$. If items are independent one has a classical test theory model for the sum score even if variances of the error term of single items are not the same. Also if the CCT model is assumed for single items there is no separation of person item parameters. In the following we consider an approach that explicitly uses item parameters.

7.3 Thresholds Models and Classical Test Theory

Let us again consider a test with several items with Y_{pi} denoting the response of person p on item i. Each item is considered a separate measurement, which means the scores are given by $X_{pi} = Y_{pi}, i = 1, \ldots, I$. Let now the thresholds model (Chap. 6) hold, that is,

$$P(Y_{pi} > y) = F(\alpha_i(\theta_p - \delta_i(y))), \quad y \in S,$$

where $\delta_i(.)$ is a non-decreasing item difficulty function defined on the support S. In the thresholds model the item function is item-specific, it represents characteristics of the item, and in its parameterized form it contains the item parameters. Thus the model starts with a separation of person and item characteristics.

At the person level the same decomposition holds as in classical test theory,

$$Y_{pi} = \mu_{pi} + \varepsilon_{pi},$$

where $\mu_{pi} = \mathrm{E}(Y_{pi})$ is the expected response and $\varepsilon_{pi} = Y_{pi} - \mu_{pi}$ is the noise variable with expectation $\mathrm{E}(\varepsilon_{pi}) = 0$. The expectation μ_{pi} represents the true score and is a function of θ_p and the difficulty function. It is implicitly assumed that the expectation is a sensible measure, which holds if responses are binary or measured on an interval scale.

For a randomly selected test taker, one obtains the decomposition

$$Y_{*i} = T_{*i} + E_{*i}, \tag{7.8}$$

where $T_{*i} = T_{*i}(\theta_p) := \mathbb{E}(Y_{*i}|\theta_p)$ and $E_{*i} = E_{*i}(\theta_p) := Y_{*i} - T_{*i}(\theta)$ are functions of the random variable θ_p. For the model (7.8) the assumptions of the CTT model (7.2)–(7.5) hold, even if the TM model does not use linear difficulty functions. (Tutz and Jordan, 2024).

The representation (7.8) has several consequences. Given that a threshold model holds a CTT model holds for randomly selected test takers. Thus, all CTT based quantities, like reliability coefficient, may be defined appropriately and all of the derived results for true score prediction may be applied to the TM setting, see also Holland and Hoskens (2003)). All the already established results derived for CTT become applicable. An important aspect is that the threshold model can be seen as a latent trait model underlying the CTT model. While in the CTT model expectations of item responses are simply considered as representing the true scores, and latent traits as the driving force behind an individual's responses on items are not clearly identified, the background model provides this identification.

When sampling from a population the distributions of the involved random variables are determined by the distribution of the corresponding characteristics in the population. Thus, the distributions of the random variables Y_{*i}, T_{*i}, E_{*i} are determined by the distribution of θ_p. If the response function is normal and the difficulty functions are linear, a normal distribution of θ_p yields normally distributed scores Y_{*i}.

In general, if a threshold model holds the distribution of Y_{*i} follows from the assumed response and difficulty functions and the distribution of the abilities. In classical test theory typically one assumes a distribution for the observed score Y_{*i} instead of assuming a distribution for single responses.

Linear Thresholds Models and the Assumptions on Distributions

The CCT model derived from a thresholds model becomes particularly simply structured if difficulty functions are linear. Thus, let us consider the linear thresholds model TM(F,linear), for which difficulty functions have the form $\delta_i(y) = \delta_{0i} + \delta_i y$, and $F(.)$ is a strictly increasing distribution function. Then, the expectation and the variance of the responses are given by

$$\mu_{pi} = \frac{\alpha_i \theta_p - \tilde{\delta}_{0i} - \mu_F}{\tilde{\delta}_i}, \quad \sigma_{pi}^2 = \frac{\sigma_F^2}{\tilde{\delta}_i^2}, \tag{7.9}$$

with item-specific parameters $\tilde{\delta}_i = \alpha_i \delta_i$, $\tilde{\delta}_{0i} = \alpha_i \delta_{0i}$, and μ_F, σ_F being constants linked to $F(.)$. The densities of (continuous) responses are given by

$$f_{pi}(y) = f(\alpha_i \theta_p - \tilde{\delta}_{0i} - \tilde{\delta}_i y))\tilde{\delta}_i,$$

where $f(.) = \partial F_{pi}(y)/\partial y$. An especially simple version is obtained in the *homogeneous* case, in which $\alpha_i = 1$, $\tilde{\delta}_i = \tilde{\delta}$. By using the parameterization $\theta_p^* = \theta_p/\tilde{\delta}$, $\kappa_i = (\tilde{\delta}_{0i} + \mu_F)/\tilde{\delta}$ one obtains

$$\mu_{pi} = \theta_p^* - \kappa_i, \quad \sigma_{pi}^2 = \sigma_F^2/\tilde{\delta}^2.$$

The corresponding density is $f_{pi}(y) = f(\theta_p - \tilde{\delta}_{0i} - \tilde{\delta} y))\tilde{\delta} = f(\tilde{\delta}(\theta_p^* - \kappa_i - y) + \mu_F)\tilde{\delta}$. That means that the mean of the response is simply the difference between the (scaled) ability θ_p^* and item difficulty κ_i rendering a simple interpretation for these parameters. In addition, the response distributions have the same form, and are just shifted versions of a basic distribution that has density $f(-\tilde{\delta} y)\tilde{\delta}$. That also means that variances are the same for all responses.

In particular the widely used CCT model with normal distributions is a special case. However, any strictly increasing function can be used as $F(.)$, which means responses can also be assumed to follow skewed distributions like the minimal or maximal extreme value distributions. Still there is a simple link between the latent ability and the item difficulty given as a simple difference.

For the CCT model derived from a thresholds model with linear difficulty function one obtains from (7.9)

$$T_i(\theta) = \mathbb{E}(Y_i|\theta) = \frac{\theta - \delta_{0i} - \mu_F/\alpha_i}{\delta_i} \tag{7.10}$$

and therefore

$$\theta = \delta_i T_i(\theta) + \delta_{0i} + \mu_F/\alpha_i.$$

It means that the resulting measurement is congeneric. Measurements are called *congeneric* if all true scores may be expressed as affine functions of a single true score, i.e., $T_i = a_i T + b_i$ holds for some fixed values a_i, b_i (Raykov, 1997). They form an important subclass of measurement models since congeneric measurement equals the notion of unidimensionality from a CTT point of view providing also the decomposition of the covariance matrix according to a one-dimensional factor analysis model.

Congeneric measurement models are called *essentially tau-equivalent* if $a_i = 1$ holds for all i in the congeneric relationships $T_i = a_i T + b_i$ (or upon redefinition of T one may just demand $a_i = c$ for some constant c). As is seen from (7.10) the model is congeneric model with $a_i = 1/\delta_i$ and $b_i = -(\mu_F \alpha_i^{-1} + \delta_{i,0})/\delta_i$. It is essentially tau-equivalent if $\delta_i = 1$ holds for all i. Tau-equivalency has some consequences: Conditionally on θ, the expected value of the unweighted mean of the item scores equals the latent variable plus a bias value (determined by the intercept term). As the bias term is independent of θ, one may deduce that the difference of means of two test takers provides an unbiased estimate of their true difference in the ability—thus justifying the usage of simple sum scores. Commonly applied coefficients—such as Cronbach's alpha—become reasonable estimators of the test reliability when tau-equivalency can be assumed (Jackson and Agunwamba, 1977).

The even stricter requirement of *essentially parallel* measurements demands in addition to tau-equivalency the equality of the error variances. It can be shown that if $\delta_i = \alpha_i = 1$ holds the condition is fulfilled (Tutz and Jordan, 2024).

In all homogeneous models one has for the mean of the sum score $Y_p = \sum_{i=1}^{I} Y_{pi}$

$$\mathrm{E}(Y_p) = I\theta_p^* - \kappa,$$

where $\kappa = \sum_i \kappa_i$ is the overall difficulty of the test. In the *non-homogeneous* case with varying discrimination parameters and slopes models are obtained, in which the variances vary across items. Then, the densities are not only shifted but also *scaled* versions of a basic distribution. However, the mean of the sum score still has a linear form

$$\mathrm{E}(Y_p) = a\theta_p - \kappa,$$

where $a = \sum_i \alpha_i/\delta_i$, and the overal difficulty is $\kappa = \sum_i (\tilde{\delta}_{0i} + \mu_F)/\delta_i$.

The thresholds model can be seen as a background model for classical test theory. In contrast to the CTT approach the thresholds model is a genuine latent trait model, which on the individual level distinguishes between the latent trait and the response as a random variable. No assumption on the distribution of person parameters is needed. This is quite different in the CTT approach, where the distribution of 'true' scores generates the randomness of responses. For further concepts of CTT derived from an underlying thresholds model see Tutz and Jordan (2024).

Non-linear Thresholds Models

Linear difficulty functions are useful for continuous responses without restrictions on the support. If the support is restricted, for example if it is strictly positive, non-linear functions are more appropriate. Within the framework of thresholds models restrictions on the support of responses are obtained in a natural way by specifying appropriate non-linear difficulty functions. This leads to models in which the mean and other characteristics of the responses are non-linear functions of the latent trait. A simple form of the difficulty function is given by $\delta_i(y) = \delta_{0i} + \delta_i g(y)$, where $g(.)$ is a strictly increasing fixed function. As example we consider responses that can only be positive values. This is obtained if for the difficulty function $\lim_{y \to 0} \delta_i(y) = -\infty$ holds, which is fulfilled if $g(.)$ is the logarithmic function, $g(y) = \log(y)$, yielding $\delta_i(y) = \delta_{0i} + \delta_i \log(y)$.

Expectations and variances of responses are no longer linear functions of the person parameter. For the logarithmic function one obtains

$$E(Y_{pi}) = c_i \exp\left(\frac{\theta_p - \delta_{0i}}{\delta_i}\right),$$

where c_i is a constant that depends on α_i, δ_i. The form of the expectation has the consequence that the total score $Y_p = \sum_i Y_{pi}$, which is often used to measure the ability, is not appropriate. One has

$$\mathrm{E}(Y_p) = \sum_{i=1}^{I} c_i \exp\left(\frac{\theta_p - \delta_{0i}}{\delta_i}\right) = \sum_{i=1}^{I} c_i \exp(-\delta_{0i}/\delta_i) \exp(\theta_p)^{1/\delta_i},$$

which is a weighted sum of exponential terms with the terms depending on the item. In the present case a condition is that items are homogeneous, that is, $\delta_i = \delta_i$ for all i. In this case one obtains

$$\mathrm{E}(Y_{p+}) = \sum_{i=1}^{I} c_i \exp(-\delta_{0i}/\delta) \exp(\theta_p/\delta) = c \exp(\tilde{\theta}_p),$$

where $\tilde{\theta}_p = \theta_p/\delta$ is the scaled ability and $c = \sum_i c_i \exp(-\delta_{0i}/\delta)$. Thus, in the homogeneous case $\mathrm{E}(Y_{p+})$ is an exponential function of the scaled ability $\tilde{\theta}_p$. Of course, one can also consider the transformed ability $\exp(\tilde{\theta}_p)$ as a measure of ability. Then,

$E(Y_{p+})$ depends linearly on the (transformed) ability. Thus, in the homogeneous case, but only then, the total score can be considered as representing the underlying ability.

Fitting of the thresholds model with continuous responses and difficulty functions that use a logarithmic transformation is illustrated in the next example.

Example 7.1 Sleep Study
In a sleep deprivation study, the average reaction time per day for subjects has been measured (data set sleepstudy from package lme4). On day 1 the subjects had their normal amount of sleep. Starting that night they were restricted to 3 h of sleep per night. The observations represent the average reaction time on a series of tests given each day to each subject. The 10 d are considered as items, measurements were available on 18 persons. Fitting of a threshold model with normal response function and logarithmic difficulty function yields log-likelihoods -882.691 (common slopes) and −863.407 (varying slopes), which means that varying slopes yield a much better fit. Figure 7.1 shoes the fitted densities for days 1, 3, 5, 8, 9; first row shows the fits for the model with common slopes, second row the fits for varying slopes. While the densities are approximately shifted versions if common slopes are assumed the distributions have quite different forms in the more flexible model with varying slopes. □

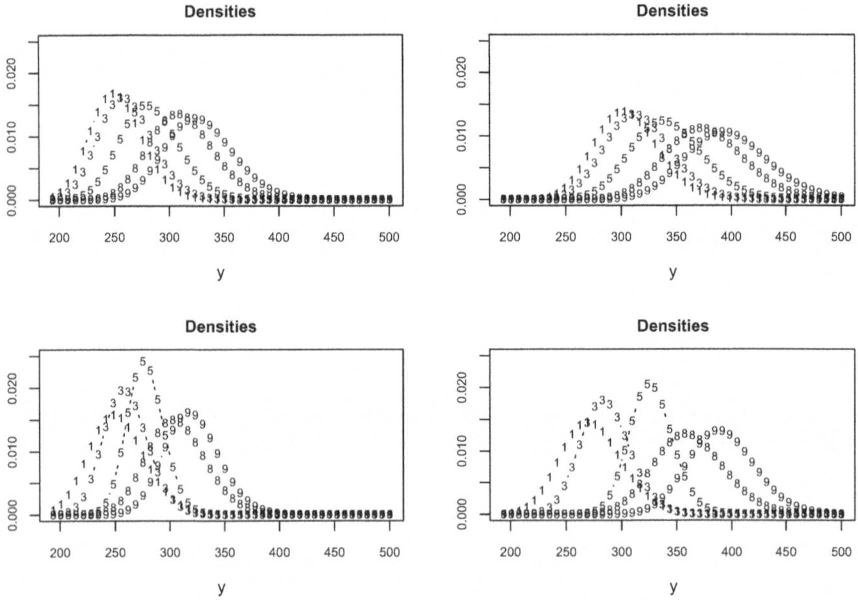

Fig. 7.1 Estimated densities for sleep study, normal response function, logarithmic difficulty function, left: $\theta = 0$, right: $\theta = 2$; first row: common slope, second row: varying slope

7.3.1 *Classical Test Theory and the Sum Score*

The focus of this book is on IRT, with classical test theory (CTT) considered only briefly, although the latter is still widely used. Nevertheless, some concluding remarks on the ongoing discussion regarding the usefulness of CTT and the sum score seem appropriate.

A distinct feature of the basic classical test theory model is that the sum score—i.e., the sum of a person's scores on the items—represents the "measurement." The sum score is highly attractive since it is easy to compute and communicate. It can be calculated for any set of items, regardless of content, and is considered the "measurement" without explicitly referring to a latent trait. Consequently, the use of the sum score has been criticized (see, for example, McNeish and Wolf (2020)). Braun and Mislevy (2005) use the term Intuitive Test Theory to describe cases where sum scores are used without careful reasoning and question "how much credence is typically attached to a single test score." In contrast, Sijtsma et al. (2024) recently made a strong case for the sum score, particularly emphasizing that it provides a good approximation in the case of binary items, even when the items follow a 2PL model. In the 1PL or Rasch model, the sum score is a sufficient statistic and, therefore, a natural indicator of performance. Since the sum score and the latent variable have a monotone likelihood ratio, respondents can still be meaningfully ordered even if the Rasch model does not hold. However, while this applies to binary items, it does not necessarily hold for Likert-type or count-type items. Moreover, order information may be less reliable when test results are used in subsequent analyses that require interval-scale measurements.

A key strength of the sum score in CTT is its simplicity and ease of use. This is why it remains widely applied, often "without accompanying evidence that scores are necessarily meaningful," as McNeish (2024) pointed out. The problem is that it is frequently used in an ad hoc manner, with little regard for validity or a clear conceptualization of the construct being measured. In its basic form, CTT relies on very weak assumptions; however, to make it a useful measurement concept, it typically needs to be linked to IRT models, for example, through the motivation provided by monotone likelihood ratios. But when referring to latent trait models, stronger assumptions must be made. These points, among others, were discussed in detail by McNeish (2024) in response to Sijtsma et al. (2024).

Despite the merits of the sum score, it seems prudent to use it with caution and to rely on its connection to latent trait models, which are necessary to complement CTT. Furthermore, more sophisticated concepts—such as multidimensionality or the inclusion of response styles—are difficult to accommodate within the CTT framework.

Chapter 8
Response Models for Count Data

Count data can be very useful in measuring abilities. Tasks that yield quite naturally count data are memory tasks where the number of remembered items is counted, divergent thinking tasks where the number of generated ideas is of interest or verbal fluency tasks where participants name instances of broad classes like animals within a short time limit, see, for example, Jansen (1995), Silvia et al. (2013), Süss et al. (2002), Doebler and Holling (2016), Forthmann et al. (2017), Forthmann et al. (2020), Forthmann and Doebler (2021).

In many assessment contexts responses are scored as non-negative integers. Nevertheless, count data as the Rasch model for count data seem not to have been widely applied. As Forthmann et al. (2020) noted, this is surprising because the RPCM can be considered as the historically first IRT model and Rasch's famous book (Rasch, 1960) was strongly concerned with the Poisson model and its potential to model reading errors.

In the following first models are considered that assume a fixed distribution for the counts. The corresponding models are the Poisson count model, the negative binomial model, and the Conway-Maxwell-Poisson model. Then a count version of the thresholds model is considered. In this model the distribution of the responses follow from a combination of response functions and difficulty functions.

In this chapter let the responses of person p on item i be given by $Y_{pi} \in \{0, 1, 2, \dots\}$ ($p \in \{1, \dots, P\}, i \in \{1, \dots, I\}$). It is assumed throughout the chapter that large counts indicate good performance, higher scores are better than lower ones.

G. Tutz, *A Short Guide to Item Response Theory Models*, Statistics for Social and Behavioral Sciences, https://doi.org/10.1007/978-3-031-87271-6_8

8.1 Rasch's Poisson Count Model

Item response models for count data date back at least to Rasch's early models (Rasch, 1960). In additive parameterization Rasch's Poisson count model specifies the expected response μ_{pi} for person p on item i by

$$\mu_{pi} = \mathrm{E}(Y_{pi}) = \exp(\theta_p - \delta_i),$$

where θ_p is the person parameter and δ_i an item parameter. For the distribution of responses a Poisson distribution is assumed with probability density function

$$P(Y_{pi} = y_{pi}|\theta_p, \delta_i) = \frac{\mu_{pi}^{y_{pi}}}{y_{pi}!}\exp(-\mu_{pi}).$$

Rasch's Poisson Count Model

$$P(Y_{pi} = y_{pi}|\theta_p, \delta_i) = \frac{\exp(\theta_p - \delta_i)^{y_{pi}}}{y_{pi}!}\exp(-\exp(\theta_p - \delta_i)) \tag{8.1}$$

The person parameter can be interpreted as ability, if θ_p increases the expected number of counts increases. The item parameter has the inverse effect, if δ_i increases the expected number of counts decreases. Thus, it is quite natural to interpret δ_i as the item difficulty. In alternative parameterizations that are in common use the difficulty is replaced by the easiness, $\tilde{\delta}_i = -\delta_i$, and the expected number increases with the easiness of the item. Since the Poisson distribution is assumed to hold for all responses the whole distribution of Y_{pi} is determined by the difference between ability and difficulty, $\theta_p - \delta_i$. Figure 8.1 shows the response distributions for three persons with abilities 3, 4, 4.5. In the left picture the item difficulty is $\delta_i = 2$, in the right picture the item is somewhat harder ($\delta_i = 2.5$). Consequently the distributions for all three persons are concentrated more closely to zero.

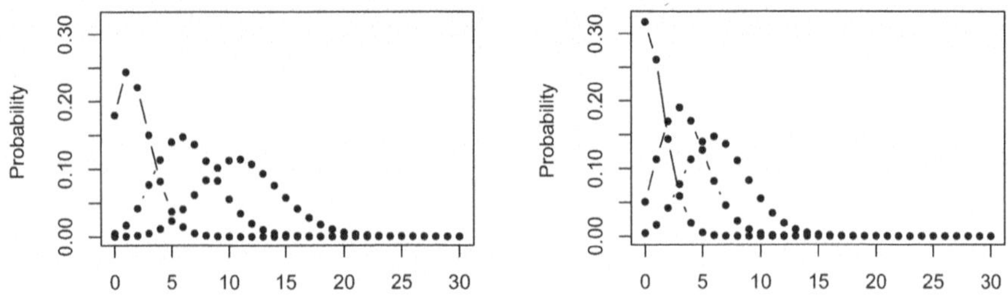

Fig. 8.1 Response probabilities for three persons with abilities 3, 4, 4.5 for item with difficulty $\delta_i = 2$ (left) and $\delta_i = 2.5$ (right)

Table 8.1 Verbal fluency data

Item	Min.	1st Qu.	Median	Mean	3rd Qu.	Max.
Animals	4.00	13.00	16.00	16.19	19.00	29.00
Letter f	5.00	9.75	11.00	11.74	14.00	21.00
Letter s	7.00	11.00	14.00	13.71	16.00	24.00
Supermarket	4.00	11.00	14.00	14.12	18.00	27.00

The assumption of a Poisson distribution is rather restrictive. In particular it implies the so-called equidispersion of Poisson distributions, which means that expected value and variance are equal for all persons and items,

$$\mathrm{E}(Y_{pi}) = \mathrm{var}(Y_{pi}).$$

Although it is quite natural that for responses for which $Y_{pi} \geq 0$ holds larger means come with larger variance, equidispersion means that they are strictly equal.

Example 8.1 Verbal Fluency
Forthmann et al. (2020) used a data set with four commonly used verbal fluency tasks (available at https://osf.io/38zsm/). The data set includes two semantic fluency tasks, namely animal naming (item 1) and naming things that can be found in a supermarket (item 4) and two letter fluency tasks, words beginning with letter f (item 2) or letter s (item 3). The participants had one minute to complete each of the verbal fluency tasks. In our analysis we use the 192 individuals that responded to all items. Table 8.1 shows some descriptive measures for the marginal distribution of item responses. The largest means are found for item 1 and item 4, which suggests that the items are easier than items 2 and 3. □

The Poisson data can be fitted by using marginal estimation with a Guassian mixing distribution by using the package *glmmTMB* (Forthmann et al., 2020).

```
library(glmmTMB)
FluencyLong<-readRDS("FluencyLong")
# Fit Poisson model
Poisson <-
  glmmTMB(value ~ -1 + item + (1 | subject_id),
  data =FluencyLong, family = poisson)
summary(Poisson)
```

The resulting output contains the easiness parameters $-\delta_i$ instead of the difficulty parameters. It is seen that the animals item is the easiest and the letter f item the hardest.

```
     AIC       BIC    logLik deviance df.resid
  4154.8    4178.0   -2072.4    4144.8       763

Random effects:

Conditional model:
 Groups      Name        Variance Std.Dev.
 subject_id (Intercept) 0.03782  0.1945
Number of obs: 768, groups:  subject_id, 192

Conditional model:
                Estimate Std. Error z value Pr(>|z|)
itemanimals      2.76595    0.02287  120.92   <2e-16 ***
itemletter f     2.44480    0.02540   96.26   <2e-16 ***
itemletter s     2.59939    0.02411  107.80   <2e-16 ***
itemsupermarket  2.62896    0.02388  110.08   <2e-16 ***
```

8.2 Negative Binomial Model

In more general models the Poisson distribution is replaced by a negative binomial distribution (Hung, 2012). Let the expected response μ_{pi} again be given by $\mu_{pi} = \mathrm{E}(Y_{pi}) = \exp(\theta_p - \delta_i)$. If the responses follow a negative binomial distribution one assumes

$$P(Y_{pi} = y_{pi}|\theta_p, \delta_i) = \frac{\Gamma(y + \mu_{pi})}{\Gamma(y+1)\Gamma(\mu_{pi})}\left(\frac{\mu}{\mu_{pi}+\nu_i}\right)^y \left(\frac{\nu}{\mu_{pi}+\nu_i}\right)^{\nu}_i, \quad y = 0, 1, \ldots$$

where $\nu_i > 0$ and $\Gamma(\nu) = \int_0^\infty t^{\nu-1}e^{-t}dt$. For the expectation and the variance one obtains

$$E(Y_{pi}) = \mu_{pi}, \quad \mathrm{var}(Y_{pi}) = \mu_{pi} + \mu_{pi}^2/\nu_i.$$

The additional parameter ν_i accounts for potential *overdispersion*. If $\nu_i > 0$ the variance of Y_{pi} is larger than the mean. In the extreme case ($\nu_i \longrightarrow \infty$) the model is equivalent to Rasch's Poisson count model. A simplified version is obtained by assuming that the additional parameter does not depend on the item, that is, $\nu_i = \nu$ for all items. The version of the negative binomial model considered here is also known as the NB2 version (Hilbe, 2011).

8.3 Conway-Maxwell-Poisson Model

The negative binomial model accounts for overdispersion but not for *underdispersion*, which occurs if the variance of Y_{pi} is smaller than the mean. A model that can account for overdispersion and underdispersion is the Conway-Maxwell-Poisson model proposed by Forthmann et al. (2020), which is an alternative generalization of the Poisson Model. Assuming conditional independence, the distribution of responses is assumed to be given by

$$P(Y_{pi} = y_{pi}|\theta_p, \delta_i) = \frac{\lambda(\mu_{pi}, \nu_i)^{y_{pi}}}{(y_{pi}!)^{\nu_i}} \frac{1}{Z(\lambda(\mu_{pi}, \nu_i), \nu_i)},$$

where $\lambda(\mu_{pi}, \nu_i)$ is the solution to $0 = \sum_{r=0}^{\infty}(r - \mu_{pi})\lambda^r/(r!)^{\nu_i}$ and $Z(\lambda(\mu_{pi}, \nu_i), \nu_i)$ is a normalizing constant, for more details of the distribution see Shmueli et al. (2005); Huang (2017).

The model contains for each item one more parameter than the Poisson model, ν_i, which has a different meaning than the corresponding parameter in the negative binomial model. The strength of the model is that it can account for overdispersion and underdispersion relative to the Poisson model. If $\nu_i < 1$, the data display overdispersion, if $\nu_i > 1$, the data display underdispersion. For $\nu_i = 1$ the item response is Poisson distributed. Thus if $\nu_1 = \cdots = \nu_I = 1$ the model simplifies to the Poisson model. It should be distinguished between the model with global dispersion parameter, $\nu_i = \nu$ for all items, and the general model with item-specific dispersion parameters.

Example 8.2 Verbal Fluency
Let us again consider the fluency data from the previous example. Fits of the Conway-Maxwell-Poisson Model with global and item-specific dispersion parameter can be obtained by using the package *glmmTMB* (Forthmann et al., 2020).

```
### estimate CMPCM with global dispersion parameter

Global <- glmmTMB(value ~ -1 + item + (1 | subject_id),
   data = FluencyLong, family = compois, dispformula = ~ 1)
summary(Global)


### estimate CMPCM with item-specific dispersion parameters

Disp <- glmmTMB(value ~ -1 + item + (1 | subject_id),
   data = FluencyLong, family = compois, dispformula = ~ -1 + item)
```

For the model with global dispersion one obtains the following output. Item parameter estimates are not given since they are virtually the same as for the Poisson model.

```
      AIC       BIC    logLik deviance df.resid
   4076.6    4104.5   -2032.3   4064.6      762

Conditional model:
 Groups     Name        Variance Std.Dev.
 subject_id (Intercept) 0.04741  0.2177
Number of obs: 768, groups:  subject_id, 192

Overdispersion parameter for compois family (): 0.553
```

The overdispersion parameter for the compois family in the output refers to $1/\nu$. Therefore, $\hat{\nu} = 1.808$ indicates underdispersion. For the model with item-specific dispersion one obtains the following output. Item parameter estimates are again almost the same as for the Poisson model.

```
     AIC       BIC    logLik deviance df.resid
  4040.1    4081.9   -2011.1   4022.1      759

Conditional model:
 Groups     Name        Variance Std.Dev.
 subject_id (Intercept) 0.0448   0.2117
Number of obs: 768, groups:  subject_id, 192

Conditional model:
                Estimate Std. Error z value Pr(>|z|)
itemanimals      2.76272    0.02089   132.3   <2e-16 ***
itemletter f     2.44153    0.01985   123.0   <2e-16 ***
itemletter s     2.59612    0.01917   135.4   <2e-16 ***
itemsupermarket  2.62572    0.02441   107.5   <2e-16 ***

Dispersion model:
                Estimate Std. Error z value Pr(>|z|)
itemanimals     -0.48495    0.12856  -3.772 0.000162 ***
itemletter f    -1.06013    0.13764  -7.702 1.34e-14 ***
itemletter s    -1.08370    0.14770  -7.337 2.18e-13 ***
itemsupermarket -0.02435    0.11733  -0.208 0.835578
```

It is seen that the model with item-specific dispersion fits much better than the simpler model. Also when accounting for the number of parameters the model dominates, AIC and BIC are smaller for the model with item-specific dispersion than in the model with global parameter. For more details on the fitting of Conway-Maxwell-Poisson models and comparison of models see Forthmann et al. (2020). □

8.4 Count Thresholds Model

The count thresholds model is the thresholds model considered in Chap. 6 applied to count data.

Count Thresholds Model

$$P(Y_{pi} > y|\theta_p, \alpha_i, \delta_i(.)) = F(\alpha_i(\theta_p - \delta_i(y))), \quad y \in \{0, 1, \ldots\}. \tag{8.2}$$

In thresholds models responses are determined by the response function $F(.)$, which is a strictly monotone distribution function and the difficulty function $\delta_i(y)$, which crucially determines the shape of the response distribution. It is assumed that for the discrimination parameter $\alpha_i \geq 0$ holds and that difficulty functions are non-decreasing. The parameters θ_p represent abilities, while α_i and $\delta_i(.)$ represent the difficulty of item i. A simplified version of the model with $\alpha_i = 1$ has been considered by Tutz (2022a).

For the difficulty function one can assume the parametric form $\delta_i(y) = \delta_{i0} + \delta_i g(y)$, where $g(.)$ is a strictly increasing function for $y \geq 0$. A simple function that suggests itself is the logarithmic difficulty function (with a shifting), which is given by

$$\delta_i(y) = \delta_{i0} + \delta_i \log(1 + y).$$

It is distinguished between models with a *common slope*, that is, $\delta_i = \delta$ for all items and the general model where slopes vary across items. The former model, if it holds, has the advantage that items are simply ordered in terms of difficulty.

As the Conway-Maxwell-Poisson model, the thresholds model can account for overdispersion and underdispersion. It does not assume that the means and variances of responses are equal. If the difficulty functions are fixed by choosing a function $g(.)$ the model is comparable to the Conway-Maxwell-Poisson model in terms of complexity. The model with common slopes contains as many parameters as the Conway-Maxwell-Poisson model with global dispersion parameters, the model with varying slopes as many parameters as the Conway-Maxwell-Poisson model with varying dispersion parameters. If difficulty functions are approximated by basis functions (Chap. 6) the thresholds models become more flexible but the number of parameters increases strongly.

Count Thresholds Model and Binary Models

There is an interesting link between thresholds models and binary models, which allows a different view on count threshold models. If the thresholds model holds one has for a fixed threshold y_0, $P(Y_{pi} > y_0|\theta_p, \delta_i(.)) = F(\alpha_i(\theta_p - \delta_i(y)))$. Let a binary response be defined by $Y_{pi}(y_0) = 1$ if $Y_{pi} > y_0$ and $Y_{pi}(y_0) = 0$ otherwise. For this variable the binary item response model

$$P(Y_{pi}(y_0) = 1|\theta_p, \gamma) = F(\alpha_i(\theta_p - \gamma_{y_0})), \tag{8.3}$$

where $\gamma_{y_0} = \delta_i(y_0)$, holds. Thus, for the collapsed binary version $Y_{pi}(y_0)$ of Y_{pi} a familiar binary model holds. The interesting point is that this holds for any threshold y_0. One consequence is that the model can also be constructed from binary models. If binary models $P(Y_{pi}(y_0) = 1|\theta_p, \delta_i(y_0)) = F(\theta_p - \delta_i(y_0))$ hold for all thresholds $y_0 \in \{0, 1, 2, \dots\}$, then the thresholds model holds. Thus, the threshold model can also be understood as a collection of binary models that have to hold simultaneously for fixed θ_p. In this sense it can be seen as an extension of binary models to count data, and binary models as special cases if count data are dichotomized. In particular, if $\alpha_i = 1$ and $F(.)$ is the logistic distribution the thresholds model is a collection of Rasch models. The link between binary models and the count thresholds model is used in the taxonomy of models given in Fig. 1.2.

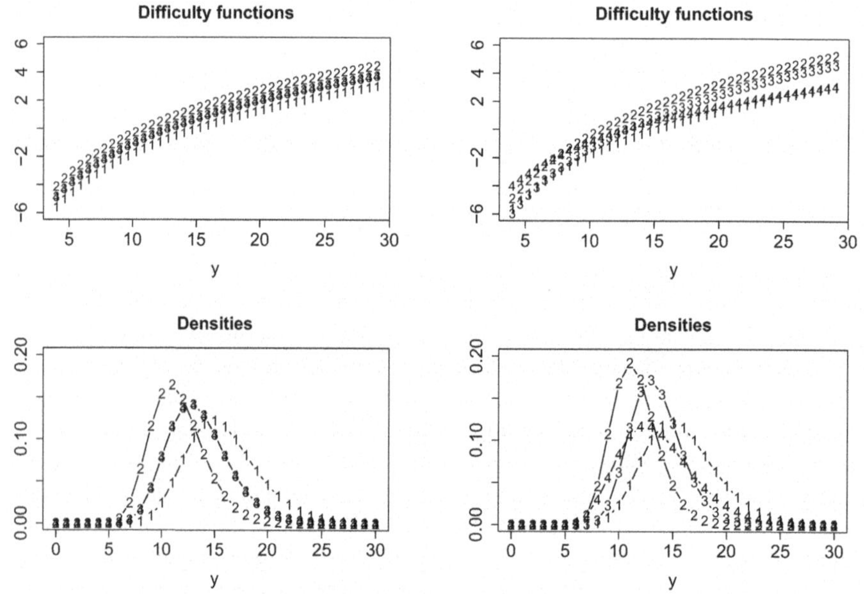

Fig. 8.2 Difficulty functions (first row) and densities (second row) for value $\theta = 0$ for verbal fluency data; left: common slopes, right: varying slopes

Example 8.3 Verbal Fluency (Tutz, 2022a)
Let us again consider the fluency data from the previous examples. Figure 8.2 shows the estimated difficulty functions and densities for value $\theta = 0$ when the normal response function is used (fixed discrimination parameters). The left column shows the fits of the model with common slopes, the right column shows the results for varying slopes. The rather restrictive model with common slopes suggests that items 3 and 4 have the same difficulty and yield identical conditional distributions. The more flexible model shows differing conditional distributions especially for these two items. Although the peak is the same for both items item 4 shows a much larger dispersion than item 3. The dispersion seems to vary strongly across items. It turns out that the more general model shows a significantly better fit and the hypotheses of common slopes should be rejected. The corresponding log-likelihood test is 54.66 on 3 df. The extended model with varying discrimination parameters showed no significant improvement (log-likelihood −2037.071) over the model with fixed discrimination parameters (varying slopes assumed).

In Table 8.2, the fits of the various models are brought together. The first part shows the results for three models, the Poisson Rasch model (M), the Conway-Maxwell model with global dispersion (CMG), and the Conway-Maxwell model with varying dispersion (CMV), which were fitted by using the R package *glmmTMB* (Forthmann et al., 2020). The parameter $\hat{\sigma}_\theta$ is the estimated standard deviation of θ obtained when using marginal likelihood estimation with normally distributed person parameters. It is seen that the Conway-Maxwell model with varying dispersion has the smallest log-likelihood and the smallest AIC.

The second part of Table 8.2 shows the results when fitting thresholds models with logarithmic difficulty functions ($\alpha_i = 1$). Also models with alternative response functions have been fitted. It is seen that the Gompertz distribution yields poor results

Table 8.2 Log-likelihoods, AIC, number of parameters, and mixture standard deviations

	Log-likelihood	Param	AIC	$\hat{\sigma}_\theta$
Poisson model	−2072.4	5	4154.8	0.194
Conway-Maxwell model, global dispersion	−2032.3	6	4076.6	0.217
Conway-Maxwell model, varying dispersion	−2011.1	9	4040.1	0.211
Logarithmic diff function, common slopes, resp fct normal	−2065.39	6	4142.79	1.041
Logarithmic diff function, varying slopes, resp fct normal	−2038.06	9	4094.11	1.092
Logarithmic diff function, varying slopes, resp fct Gumbel	−2011.68	9	4041.36	1.296
Logarithmic diff function, varying slopes, resp fct Gompertz	−2095.88	9	4209.76	1.457
Eight splines, resp fct normal	−1994.75	33	4055.50	0.972
Logarithmic diff function + eight splines, resp fct normal	−1990.40	41	4062.80	0.973

but if the Gumbel distribution serves as response function the fit of the model is virtually the same as for the most general Conway-Maxwell model, which has the best fit among the fixed distribution models.

The third part of Table 8.2 shows models with data-driven difficulty functions. One model uses eight cubic B-splines to approximate the difficulty functions while the other one uses a combination of the logarithmic function and eight cubic B-splines. The difference between the two models is negligible. Also the likelihood ratio test that compares the two model shows that there is no significant difference (8.70 on 8 df). The model with a logarithmic function and B-splines has the advantage that it can be compared to the model with logarithmic difficulty functions since the latter is a submodel of the former. The likelihood ratio test yields 95.12 on 32 df, which means the difference is highly significant, and the model that allows for more flexible difficulty functions is more appropriate.

The largest log-likelihood value was found for the extended B-splines model but the AIC is larger since it has more parameters than the other models. In terms of AIC the best models are the general Conway-Maxwell model and the thresholds model with logarithmic difficulty functions and the Gumbel distribution as response function. It is worth mentioning that posterior estimates are highly correlated for the models. For example, the person parameters of the threshold model with varying slopes have correlation 0.9906 with the person parameters of the Poisson model, 0.9924 with the global dispersion Conway-Maxwell model, and 0.9901 with the varying dispersion Conway-Maxwell model. For the latter model the computational effort is rather high as compared to threshold models. □

8.5 Overview

Figure 8.3 shows the taxonomy of count IRT models. The Conway-Maxwell-Poisson model, the negative binomial model, and the counts thresholds model are not nested,

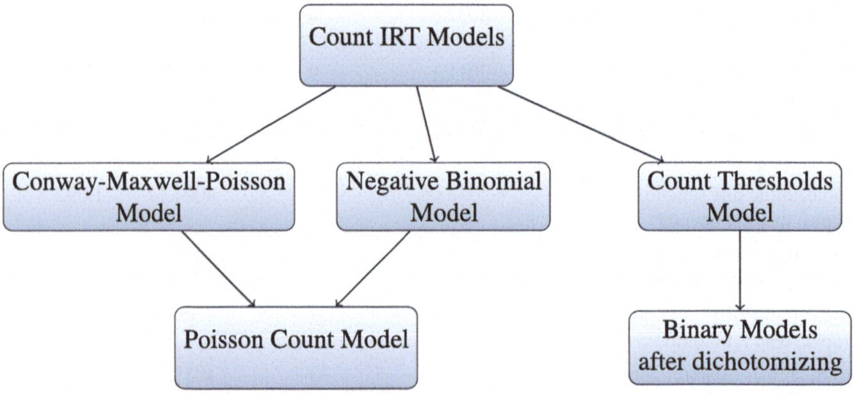

Fig. 8.3 Count data latent trait models

they represent different approaches to model item response data. The Conway-Maxwell-Poisson model as well as the negative binomial model are generalizations of the Poisson model and contain it as special case. The thresholds models contains traditional binary models as special cases if responses are dichotomized.

The Conway-Maxwell-Poisson model and the thresholds model are much more flexible than the Poisson model and the negative binomial model. The latter are often too restrictive to capture the underlying response distributions. In terms of goodness-of fit the Conway-Maxwell-Poisson model and the thresholds model with an appropriately chosen response function are well comparable. A potential advantage of the thresholds model is that it can be combined with binary and polytomous items that are assumed to measure the same latent trait. More details on the Conway-Maxwell-Poisson model are found in Forthmann et al. (2020), Forthmann and Doebler (2021), where the model is also compared to its simpler versions. More details on the count thresholds model are found in Tutz (2022a).

8.6 Fitting of Models

Fits of the Poisson Rasch model and the Conway-Maxwell model can be obtained by using the R package *glmmTMB* (Forthmann et al., 2020). The counts thresholds model can be fitted by special software available at the github repository of the book.

Chapter 9
Tree-Based Item Response Models

Tree-based item response models are hierarchically structured models that assume a nested structure with the building blocks given by traditional binary or polytomous models. Most of the models can be seen as recursive partitioning models in which the set of response categories is successively split into subsets of categories. In particular models with binary splits have been used widely yielding so-called IRTrees models (Böckenholt (2017), De Boeck and Partchev (2012), Jeon and De Boeck (2016), Böckenholt and Meiser (2017)). Though specific models are very useful in achievement tests, a great deal of the literature is concerned with the measurement of attitudes, beliefs, assessments of importance, and assessments of satisfaction (Böckenholt and Meiser, 2017). An often used format are Likert scales, in which the grades are assumed to be ordered and reflect agreement/disagreement or approval/disapproval of the respondent with respect to the value statement. Tree-based item response models for Likert items assume a sequential judgment process. One strength of the approach is that it can account for a priori specified response styles such as the tendency to extreme or moderate categories.

In the following we consider binary structured IRTrees as well as multi-way trees, in which the building blocks are multinomial, typically ordinal models. In general the construction of trees depends on the item format. For different applications quite different trees are useful but consideration of trees is restricted to only few item formats.

9.1 Binary Item Response Trees

Item response trees models (IRTrees) are sequential process models for categorical responses such that the choice of a response category can be represented by a response tree. A simple example is the tree given in Fig. 9.1. It represents the tree when items are solved in consecutive observed steps and the transition from one level (category)

G. Tutz, *A Short Guide to Item Response Theory Models*, Statistics for Social and Behavioral Sciences, https://doi.org/10.1007/978-3-031-87271-6_9

to the next is modeled conditionally given the previous level has been reached. It is the tree underlying the sequential model considered in Sect. 4.4. In other applications trees can sometimes be seen as representing a "sequential response process, according to which a rating process is constructed based on the respondent's answers to a series of mental queries that are prompted by both the item format and content" (Böckenholt, 2017). Let us consider an ordinal response with six categories representing ordinal outcomes ranging from "strongly disagree" to "strongly agree". In the tree shown in Fig. 9.2, the first query determines a respondent's agreement or disagreement. The second query determines the extremity of the (dis)agreement and the third query assesses whether the (dis)agreement is weak or not. Each query in the tree corresponds to a conditional binary decision.

The trees in Figs. 9.1 and 9.2 are partition trees. With the understanding that nodes correspond to sets of categories a directed tree for categories is called a *partition tree* if

- the root is given by the set of all response categories,
- child nodes (vertices of the graph) are obtained as a partition of the categories in the parent node,
- every response category is in the tree as a terminal node.

In this and the following section the trees are assumed to be partition trees, alternative trees will be considered briefly in Sect. 9.3.

Binary Trees and Pseudo-Items

Binary IRTrees as considered in this section have two branches from each node unless they are end nodes. The branching can refer to the ability of mastering a specific step (Fig. 9.1), to binary decisions as in attitude questionnaires (Fig. 9.2) or to other underlying processes. The essential trait is that branching corresponds to a (local) binary response, which can be considered a binary pseudo-item. Each internal node is linked to a binary *pseudo-item* while end nodes represent categories.

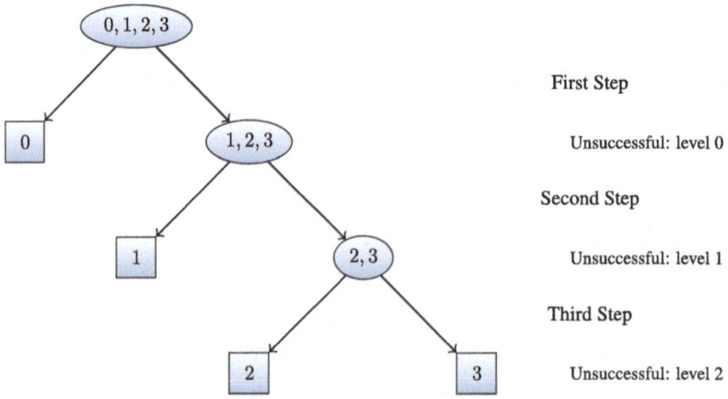

Fig. 9.1 The sequential model as a hierarchically structured model

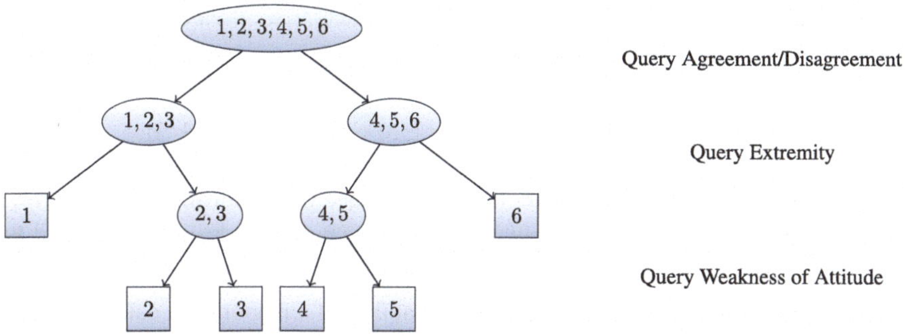

Fig. 9.2 A tree for six ordered categories, categories 1, 2, 3 represent levels of disagreement, categories 4, 5, 6 represent levels of agreement

The tree represents the sequential process that yields a final response by "solving" pseudo-items.

Let $Y_{pi}^{(q)}$ denote a binary pseudo-item that captures the outcome of query q for item i by person p. For the probability of a positive outcome a binary model is assumed to hold, which typically has the form of a traditional binary model,

$$P(Y_{pi}^{(q)} = 1) = F(\alpha_i^{(q)}\theta_p^{(q)} + \delta_i^{(q)}), \tag{9.1}$$

where $\theta_p^{(q)}$ is a node-specific person parameter, $\alpha_i^{(q)}$ is a discrimination parameter, and $\delta_i^{(q)}$ is an "easiness" parameter.

For a general description of trees, it is helpful to characterize pseudo-item responses more distinctly. Each pseudo-item or query q is linked to two sets of categories $S_1^{(q)}$, $S_2^{(q)}$, which represent the dichotomization of categories that is captured by the pseudo-item. For example, the first pseudo-item in the tree given in Fig. 9.1 is characterized by $S_1^{(q)} = \{0\}$, $S_2^{(q)} = \{1, 2, 3\}$, the second pseudo-item is characterized by $S_1^{(q)} = \{1\}$, $S_2^{(q)} = \{2, 3\}$. Let $(S_1^{(q)}, S_2^{(q)})$ denote the tuple of sets linked to pseudo-item q. Then, the pseudo-item response $Y_{pi}^{(q)}$ can be given by

$$Y_{pi}^{(q)} = 1 \text{ if } r \in S_2^{(q)} \text{ is selected,} \quad Y_{pi}^{(q)} = 0 \text{ if } r \in S_1^{(q)} \text{ is selected.}$$

It is essential that pseudo-item responses corresponding to internal nodes are *conditional* responses. They refer to the partitioning of the combined set $S^{(q)} = S_1^{(q)} \cup S_2^{(q)}$ into $S_1^{(q)}$ and $S_2^{(q)}$, that is,

$$Y_{pi}^{(q)} = 1 \text{ is equivalent to } Y_{pi} \in S_2^{(q)} \text{ given } Y_{pi} \in S^{(q)}.$$

For example, for the second pseudo-item in Fig. 9.1, which is characterized by $S_1^{(q)} = \{1\}$, $S_2^{(q)} = \{2, 3\}$ one obtains that

$$Y_{pi}^{(q)} = 1 \text{ is equivalent to } Y_{pi} \in \{2, 3\} \text{ given } Y_{pi} \in \{1, 2, 3\}.$$

Each response category r results from a sequence of pseudo-items $q_1, \ldots, q_m$, where the dependence on r is omitted. The sequence of pseudo-items is linked to a sequence of indicator functions $I_1, \ldots, I_m$ with $I_j = 1$ if $r \in S_2^{(q_j)}$ and $I_j = 0$ if $r \in S_1^{(q_j)}$. Using these two sequences the probability of a response in category r has the form

$$\begin{aligned} P(Y_{pi} = r) &= \prod_{j=1}^{m} P(Y_{pi}^{(q_j)} = 1)^{I_j} (P(Y_{pi}^{(q_j)} = 0)^{1-I_j} \\ &= \prod_{j=1}^{m} F(\alpha_i^{(q_j)} \theta_p^{(q_j)} + \delta_i^{(q_j)})^{I_j} (1 - F(\alpha_i^{(q_j)} \theta_p^{(q_j)} + \delta_i^{(q_j)}))^{1-I_j}. \end{aligned}$$

Sparsity in Parameterization

Trees as given in Figs. 9.1 and 9.2 show the *processing model*. They show in particular which pseudo-items are involved. The tree-based model itself is the parameterized model that is constructed according to the tree structure. IRTree models typically are multidimensional such that the response probabilities are determined by a vector of person parameters. In the most general case as specified in (9.1), each pseudo-item has its own node-specific person parameter. Then the person parameter $\theta_p^{(q)}$ represents the person's ability to "solve" pseudo-item q and the interpretation of the parameter refers to the specific pseudo-item. Although models of this type have been used (Plieninger, 2020), they tend to be overparameterized and ignore that there might be a common latent trait that is involved in all pseudo-items. Typically one wants sparser models with a clear interpretation of the person and item parameters that account for the ordering of responses.

The simple tree shown in Fig. 9.1 for responses $Y_{pi} \in \{0, \ldots, k\}$ is a *linear tree*, that is a tree in which one branch from each internal node leads directly to an end node. A simple parameterization of pseudo-items is obtained by using for pseudo-item q characterized by the sets $(S_1^{(q)}, S_2^{(q)}) = (\{i\}, \{i+1, \ldots, k\})$ the model

$$P(Y_{pi}^{(q)} = 1) = P(Y_{pi} \in \{i+1, \ldots, k\} | Y_{pi} \in \{i, \ldots, k\}) = F(\alpha_i^{(q)} \theta_p + \delta_i^{(q)}),$$

with the person parameter θ_p not depending on the pseudo-item. The model uses just one person parameter that represents the ability to obtain a high score. The parameterization is sparse and equivalent to the sequential model considered in Sect. 4.4. It aims at measuring the ability of a person by using polytomous items that can be solved in a step-wise manner only.

Also in non-linear trees, which are used in the measurement of attitudes, it is useful to clarify which latent trait(s) one intends to measure. The tree in Fig. 9.2 is a non-linear tree with five nodes that are not end nodes, thus being characterized by five pseudo-items. The response refers to agreement in Likert item with six categories and the basic latent trait that is to be measured is the attitude. For the first pseudo-item, which splits the six categories into two groups of categories one can assume

$$P(Y_{pi}^{(1)} = 1) = P(Y_{pi} \in \{4, 5, 6\}) = F(\alpha_i^{(1)}\theta_p^{(a)} + \delta_i^{(1)}),$$

where $\theta_p^{(a)}$ represents the person's global tendency to agree on a very coarse scale. For the pseudo-item 2, which splits $\{1, 2, 3\}$ into $\{1\}$ and $\{2, 3\}$ a model that includes the tendency to agree but also other person parameters is given by

$$P(Y_{pi}^{(2)} = 1) = P(Y_{pi} \in \{2, 3\}|Y_{pi} \in \{1, 2, 3\}) = F(\alpha_i^{(a2)}\theta_p^{(a)} + \alpha_i^{(2m)}\theta_p^{(m)} + \delta_i^{(2)}).$$

It contains the parameter $\theta_p^{(a)}$, which is the global tendency to agree, weighted by $\alpha_i^{(a2)}$, and the additional person parameter $\theta_p^{(m)}$, which represents the tendency to a moderate response weighted by $\alpha_i^{(2m)}$. If $\theta_p^{(a)}$ increases the probability to prefer categories 2, 3 over 1 increases. The same holds for $\theta_p^{(m)}$. While the former represents the general tendency to higher scores the latter is intended to represent the tendency to moderate responses, which becomes clear in the parameterization of pseudo-item 3. For pseudo-item 3, which splits $\{4, 5, 6\}$ into $\{4, 5\}$ and $\{6\}$ one can specify

$$P(Y_{pi}^{(3)} = 1) = P(Y_{pi} = 6|Y_{pi} \in \{4, 5, 6\}) = F(\alpha_i^{(a3)}\theta_p^{(a)} - \alpha_i^{(2m)}\theta_p^{(m)} + \delta_i^{(3)}),$$

which again contains $\theta_p^{(a)}$ and $\theta_p^{(m)}$, however the latter is included with a negative sign. Increasing parameter $\theta_p^{(m)}$ indicates a stronger preference for categories 4, 5, and is therefore also represents the tendency to moderate responses. Thus, in pseudo-items 2 and 3 the same parameter $\theta_p^{(m)}$ is used to represent the tendency to moderate responses. The model for the tree is constructed such that it measures the tendency to agree but accounts for the response style, which is measured by a separate parameter. The representation is rather economic. A less sparse parameterization would result if different response style parameters were used for pseudo-items 2 and 3, therefore assuming that the tendency to extreme or moderate responses differs within agreement and disagreement categories. Since the corresponding models (with and without varying response style parameter) are nested it could be tested if the sparser parameterization is sufficiently complex.

The example illustrated that it is useful to specify more than one person parameter for a given node. Integrating the target trait (agreement) in pseudo-item that come later in the tree and contain the response style seems sensible since the choice among agreement (or disagreement) categories is also a choice between low and high categories. While early tree concepts used simple node-specific person parameters sparser, more sensible models were propagated much later, for example by Meiser et al. (2019). The models used for pseudo-items 2 and 3 are bifactor models, which load on $\theta_p^{(a)}$. and $\theta_p^{(m)}$ while the pseudo-item 1 loads on $\theta_p^{(a)}$ only.

In general, parameterizations with a varying degree of complexity can be used. If models are nested one can test if a simpler parameterization is sufficiently complex. However, starting with a very complex parameterization and successively simplifying has the disadvantage that one has to fit multidimensional models, which, when using marginal estimation methods means high dimensional integration. Moreover, it cannot be assumed that person parameters are uncorrelated since, for example, the tendency to agree on a Likert item might be correlated with the tendency to moderate or extreme responses.

Tree-based latent trait models were considered, among others, by De Boeck and Partchev (2012), Thissen-Roe and Thissen (2013), Böckenholt (2012), Khorramdel and von Davier (2014), Böckenholt (2017) and Böckenholt and Meiser (2017), Meiser et al. (2019), Tutz (2020a).

9.2 More General Partition Trees

Binary trees are easy to fit and understand but have some limitations. For example, in a Likert item with five categories it is unclear how to choose the first partitioning into subsets if one wants to exploit the ordering of categories. Alternatives that suggest itself are the partitioning $\{1, 2\}, \{3, 4, 5\}$ and the partitioning $\{1, 2, 3\}, \{4, 5\}$. Both use the ordinal structure but assume different processes, none of which seems convincing as a process model for a respondent's choice of categories. Of course, the first pseudo-item can be chosen such that it separates the middle category from the rest of the categories, however, this choice does not exploit the ordering of categories. More general trees that are not restricted to binary pseudo-items as considered in the following allow for smaller trees.

Multi-way Trees

In more general trees the pseudo-items are not necessary binary but can include multi-way splits. Let us consider the tree given in Fig. 9.3. The first pseudo-item has three response categories corresponding to the partitioning $\{1, 2\},\{3\}, \{4, 5\}$. The other two pseudo-items are binary. For the first split the ordering of categories is exploited when using a classical ordinal model for three response categories, for example, the graded response model,

$$P(Y_{pi} \geq 3) = F(\theta_p - \delta_1^{(1)}), \quad P(Y_{pi} \geq 4) = F(\theta_p - \delta_2^{(1)}),$$

with thresholds $\delta_1^{(1)} \leq \delta_2^{(1)}$ and person parameter θ_p. For simplicity a model without discrimination parameter has been used, but it can also be included. For the other two binary pseudo-items, one can use

$$P(Y_{pi} = 2|Y_{pi} \in \{1, 2\}) = F(\theta_p + \theta_p^{(m)} - \delta_1^{(2)}), \quad P(Y_{pi} = 5|Y_{pi} \in \{4, 5\}) = F(\theta_p - \theta_p^{(m)} - \delta_1^{(3)}).$$

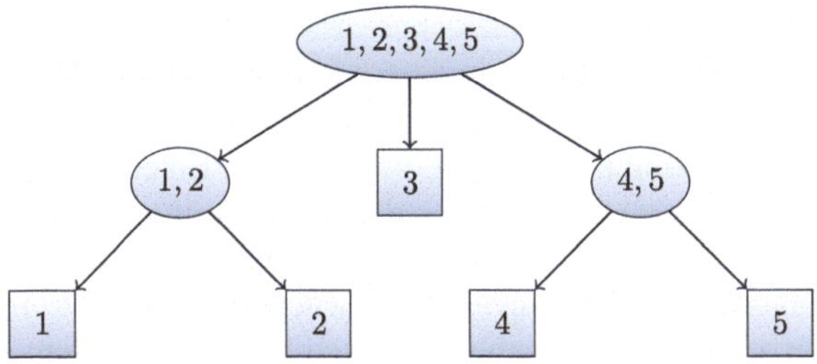

Fig. 9.3 A tree for five ordered categories with three levels

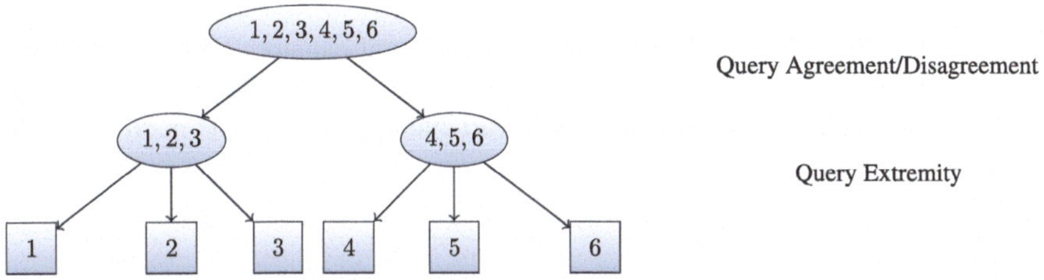

Fig. 9.4 A tree for six ordered categories with three levels

The two models contain the person parameter θ_p, which represents the person's tendency to agreement (higher categories) and the parameter $\theta_p^{(m)}$, which represents the tendency to moderate responses. Correspondingly, $-\theta_p^{(m)}$ can be interpreted as the tendency to extreme responses. In both cases, it is a response style parameter. To obtain a sparse representation, the tendency to middle or extreme responses is the same when distinguishing between 1 and 2 and distinguishing between 4 and 5.

A similar model with just three levels in the tree can be formulated for the tree given in Fig. 9.4. The first pseudo-item is binary and separates the agreement categories from the disagreement categories. The second and third pseudo-items have three response categories and can be modeled by any ordinal model including the general tendency to agree by using the same parameter as in the first pseudo-item and an additional response style parameter that represents the tendency to moderate or extreme categories. The corresponding model is simpler than the binary tree model shown in Fig. 9.2, which uses somewhat artificial pseudo-items in the query weakness of attitude. The multi-way tree model in Fig. 9.4 might much better reflect the underlying process, first deciding if to agree or not and in the second step deciding on the degree of disagreement or agreement.

Thissen-Roe and Thissen (2013) gave an application using a two-decision model that includes multi-way pseudo-items by using a modified graded response model in the second level. For the estimation one can use similar methods as in binary trees, exploiting that likelihood contributions can be written as products of conditional probabilities, see also Tutz and Draxler (2019).

Likert Items with a Midpoint

In the following, we consider briefly trees for Likert items in which the number of response categories is odd since it is a quite common response format and raises interesting questions concerning the ordering of categories. Likert scales as the five and seven-grade scales use a mid-point that indicates neutrality ("neutral") or ambivalence ("neither agree nor disagree"). The role of the neutral category is ambivalent. For example, Kulas et al. (2008) investigated whether it is used to indicate a moderate standing on a trait/item, or rather is viewed by the respondent as a "dumping ground" for unsure or non-applicable response. As illustrated by Tutz (2020a) in the latter case, the use of the middle category as part of the integer protocol might yield strongly biased results.

A hierarchical tree model that separates the neutral category and has just three levels is given in Fig. 9.5. Let $m = (k+1)/2$ denote the middle category and the first pseudo-item $Y_{pi}^{(1)}$ represent the binary variable that codes if the neutral middle category is chosen,

$$Y_{pi}^{(1)} = \begin{cases} 1 & Y_i \neq m \\ 0 & Y_i = m. \end{cases} \tag{9.2}$$

Thus, $Y_{pi}^{(1)} = 1$ indicates that the respondent has shown some degree of preference, $Y_{pi}^{(n)} = 0$ indicates that the indecision category was chosen. If $Y_{pi}^{(1)} = 1$ only categories $1, \dots, m-1, m+1, \dots, k$ can occur. The variable that indicates the degree of agreement can be given by

$$\tilde{Y}_{pi} = \begin{cases} Y_{pi} & Y_{pi} \leq m-1 \\ Y_{p,i-1} & Y_{pi} \geq m+1. \end{cases}$$

Variable $\tilde{Y}_{pi}$ is a rescaled version of Y_{pi} with the neutral category omitted, the new categories $1, \dots, m-1$ represent the disagreement categories, and $m, \dots, k-1$ the agreement categories. The tree has just two components. The first pseudo-item can be chosen by

$$P(\tilde{Y}_{pi}^{(1)} = 1) = F(\alpha_i^{(1)}(\theta_p^{(n)} - \delta_i^{(1)})), \tag{9.3}$$

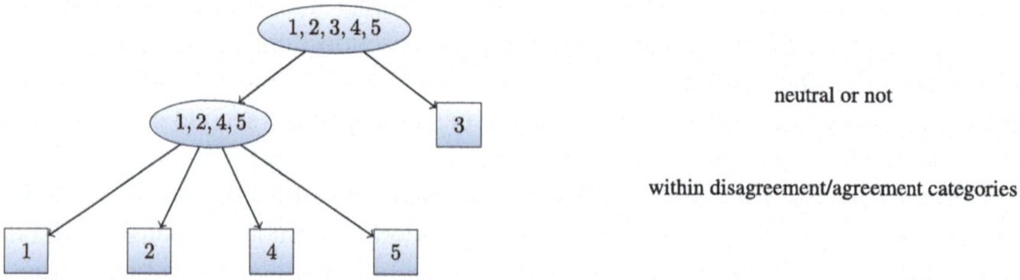

Fig. 9.5 A tree for five ordered categories, categories 1, 2 represent low response categories, categories 4, 5 represent high response categories, 3 is the neutral middle category

where $\theta_p^{(n)}$ is the person parameter that represents the tendency to choose the neutral category. The second component can be chosen as an adjacent categories model,

$$P(\tilde{Y}_{pi}^{(2)} \geq r | \tilde{Y}_{pi}^{(2)} \in \{r-1, r\}) = F(\alpha_i^{(2)}(\theta_p^{(a)} - \delta_{ir}^{(2)})), \tag{9.4}$$

where $\theta_p^{(a)}$ is the person parameter that represents the tendency to agree. The model has two person parameters, $\theta_p^{(n)}$ and $\theta_p^{(a)}$, which have a clear interpretation. Under the assumption that the person parameters are independent one can even estimate the models separately. If one is not willing to make this assumption one has to fit a two-dimensional model with potential correlations between the person parameters. The ordinal model (9.4) can also be extended to contain a response style parameter by the models given in Chap. 5.

The neutral category can also be separated by using binary trees as shown in Fig. 9.6. The first pseudo-item is the same as before, the other pseudo-items are binary. For the former one might use model (9.3), for the others the models considered in the section on binary trees can be used. In the case of five response categories it is straightforward to parameterize such that one has three parameters, one for the tendency to choose the neutral category, one for the tendency to agree, and one for the tendency to extreme or moderate responses. For seven-grade scales binary tree models become more complex since more pseudo-items are involved. Then one can use models as considered for the tree Fig. 9.2.

Probabilities in Partition Trees

Let a tree be defined by m pseudo-items $Y_{pi}^{(1)}, \ldots, Y_{pi}^{(m)}$. Let the partitioning for jth pseudo-item be given by $S_1^{(j)}, \ldots, S_{k_j}^{(j)}$, which are disjunct subsets of response categories such that the pseudo-item is given by

$$Y_{pi} \in S_l^{(j)} \text{ given } Y_{pi} \in S^{(j)},$$

where the combined set $S_j = S_1^{(j)} \cup \cdots \cup S_{k_j}^{(j)}$ is the set of categories partitioned by the jth pseudo-item.

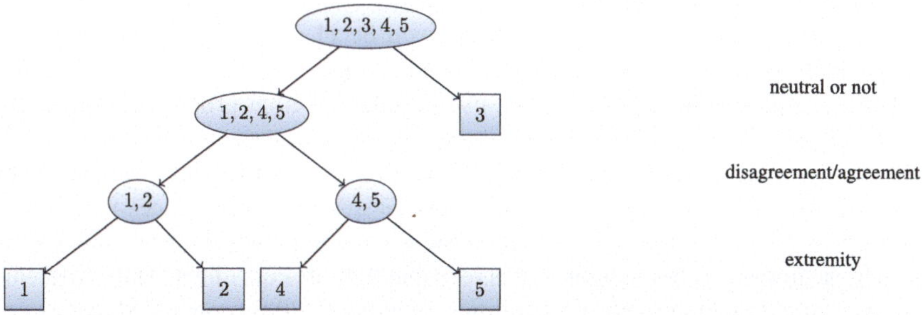

Fig. 9.6 A tree for five ordered categories, categories 1, 2 represent low response categories, categories 4, 5 represent high response categories, 3 is the neutral middle category

Then, the probability of a response in category r has the form

$$P(Y_{pi} = r) = \prod_{j=1}^{m} \prod_{l=1}^{k_j} P(Y_{pi} \in S_l^{(j)} | r \in S^{(j)})^{I(r \in S_l^{(j)})}$$

where $I(.)$ is an indicator function with $I(c) = 1$ if c is true and $I(c) = 0$ otherwise.

9.3 Alternative Trees

Most of the literature uses partition trees without making it explicit. Interesting trees that are not partition trees and raise questions of interpretation have been considered by Plieninger (2021). Let us consider a Likert item with five response categories. The tree given in Fig. 9.7 is not a partition tree since the branching of the root does not correspond to a partition of the categories (the sets in the child nodes are not disjunct). However, it can be understood as a query concerning the trait, distinguishing between low ($\{1, 2, 3\}$) and high ($\{3, 4, 5\}$) categories. Therefore, the branching can be seen as a binary pseudo-item referring to the trait, which is made explicit by including "trait t_{pi}" in the root node. The other pseudo-items involved refer to actual partitions and are rather obvious, the extreme response style pseudo-item denoted by "extreme e_{pi}" distinguishes, for example, between $\{1\}$ $\{2, 3\}$, the mid-point response style pseudo-item denoted by "mid-p m_{pi}" distinguishes, for example, between $\{2\}$ $\{3\}$.

Let the probability of the pseudo-items be given by

$$\begin{aligned} t_{pi} &= F(\alpha^{(t)}\theta_p^{(t)} + \delta^{(t)}) \text{ for the trait pseudo-items} \\ m_{pi} &= F(\alpha^{(m)}\theta_p^{(m)} + \delta^{(m)}) \text{ for the mid-point response style pseudo-items} \\ e_{pi} &= F(\alpha^{(e)}\theta_p^{(e)} + \delta^{(e)}) \text{ for the extreme response style pseudo-items} \end{aligned}$$

The resulting probabilities are given in the lower part of Fig. 9.7, for example the probability of response category 1 is given by $P(Y_{pi} = 1) = e_{pi}(1 - t_{pi})$. The probability of response category 3 is composed from two components because it can result from two pseudo-items. It is given by $(1 - e_{pi})m_{pi} = (1 - t_{pi})(1 - e_{pi})m_{pi}(\text{ left branch}) + t_{pi}(1 - e_{pi})m_{pi}(\text{ right branch})$.

The same probabilities result if one uses a quite different process tree, namely the one given in Fig. 9.8. The structure is quite different. The pseudo-item at the root splits between the extreme categories $\{1, 5\}$ and the rest. Mid-point response style pseudo-items and trait pseudo-items are found at quite different places within the tree than in the tree in Fig. 9.7. Nevertheless, when comparing the probabilities as given in the lower part of the figures, it is seen that they are equivalent. Thus different processes might yield the same probabilities. Plieninger (2021) concluded that a tree model does not imply a certain sequential order of the psychological processes. He states that " it is irrelevant for the model equations in which order the processes occur

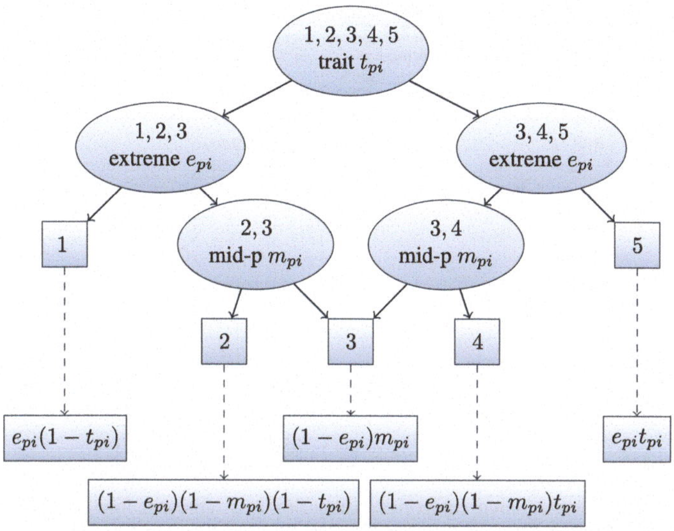

Fig. 9.7 An alternative tree for five ordered categories, categories 1, 2 represent low response categories, categories 4, 5 represent high response categories, 3 is the neutral middle category

on a path in a tree diagram, it is only relevant which processes occur. Thus, IR-tree models make assumptions about the number and type of processes involved but not about their order, and such interpretations can be misleading."

Among the two processes that yield the same model when pseudo-items are specified accordingly one is a partition tree the other is not. Indeterminacy of the process certainly occurs if one allows for trees that are not partition trees.

The trees are binary trees for a five-point item and could be compared to the process model used for the tree in Fig. 9.3. The latter is sparser parameterized using only two person parameters while the models considered in this section use three person parameters, for trait, extreme, and mid-point response style. A simpler parameterization would be obtained if the mid-point response style was replaced by the trait. For example, in the model for Fig. 9.7 m_{pi} could be replaced by $m_{pi} = F(\alpha^{(m)}\theta_p^{(t)} + \delta^{(m)})$, which uses the trait parameter $\theta_p^{(t)}$ but has own item parameters. This is warranted since the choice between 2 and 3 (or 3 and 4) is also a choice between ordered categories, which is ignored by the model. In contrast to the models considered in Sect. 9.1 the binary pseudo-models use just one person parameter. It might be appropriate to let the choice between 2 and 3 (or 3 and 4) depend on the trait parameter and a mid-point response style parameter (for 3 and 4 with a reverse sign) if one wants to account for a special mid-point response style. Also in the models for Fig. 9.8 the mid-point split ignores that the choice between 3 and $\{2, 4\}$ ignores that it might also be determined by the trait since it is a choice between two groups of categories that are ordered.

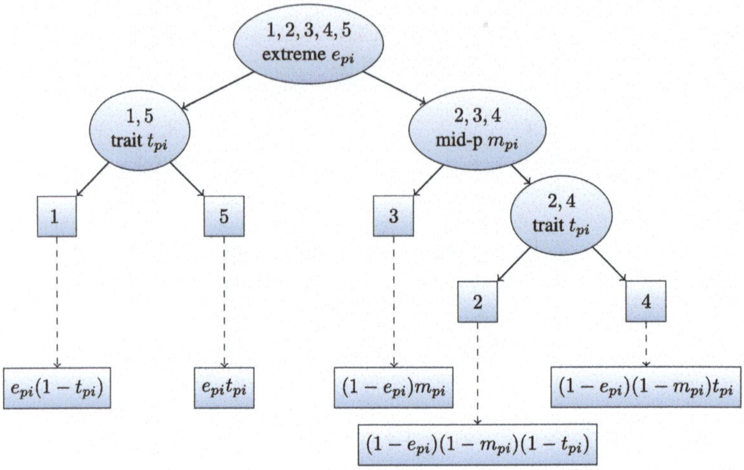

Fig. 9.8 A tree for five ordered categories, categories 1, 2 represent low response categories, categories 4, 5 represent high response categories, 3 is the neutral middle category

9.4 Modeling with Trees

The strength of tree-based item response models is that they can account for response styles, in particular the preference for extreme or middle categories. The modeling approach is a conditional one. The process that is expressed in a tree diagram assumes that categories are always chosen given a restricted set of categories. The final model is composed from conditional pseudo-items, which are the building blocks of the model. The approach is very flexible since quite different pseudo-items and corresponding models can be used. Referring back to binary or polytomous pseudo-items with a reduced number of categories has the advantage that simple models with a clear interpretation of parameters are used.

Flexibility of a modeling approach comes with the burden of having to select an appropriate model. The choice of the process model can be driven by a psychological theory, the item format or both. In some cases there is a process that suggests itself, for example, if an item in an achievement test can be solved only in a step-wise manner. Then, the sequential model is a sensible choice triggered by the item format. Also other formats like the two-stage funnel format considered, for example, by Böckenholt (2017), Meiser et al. (2019) determine the choice of the process model. In the case of Likert items the choice is more difficult. Binary trees tend to bring about overparametrized models, for example, when postulating a specific preference for mid-categories instead of a general tendency to categories in the middle. Polytomous pseudo-items typically offer simpler models with fewer layers and parameters.

Although specific process model might seem appropriate and motivated by a psychological theory it is to be emphasized that a process is just *assumed* to hold and it is not easy to validate the choice by data analysis. There are hardly formal tests that can distinguish between models. Even in the simplest case, in achievement tests that can only be solved in a step-wise manner, one might fit a sequential model and compare it to a graded response or adjacent categories model, which do not assume a sequential process. Of course one can compare the goodness-of-fit, but that depends on the parameterizations that are used and even if one accounts for the complexity by using AIC or similar measures, these measures can only give hints at which model to choose. The problem is distinctly more demanding in trees for Likert items. Although the tree and the associated sequential structure can be turned into a model by parameterizing accordingly it does not work the other way round, that is, the model does not provide a test of the sequential structure, which is obvious from the equivalent models considered for Figs. 9.7 and 9.8. In general, a more reasonable guide for the choice of models might be to have a clear concept of the features that one wants to model, for example, if one wants to investigate the steps that lead to a higher level in the performance. The search for the true generating (process) model is bound to fail.

The role of the process is further complicated by the fact that the order of the process is not fixed as shown in Sect. 9.3. Alternative trees can lead to the same probabilities but have a quite different order. Therefore, some caution is needed in the interpretation of parameters. The optimism that is seen in early publications on tree-based models, that a fixed sequential process is postulated and evaluated, seems less warranted after these findings.

Tree-models are designed for categorical responses. In most applications the response categories are at least partially ordered. A strength of the models is that they can model that all categories are ordered, for example in the trees in Figs. 9.1 and 9.2, but they can also account for partial order by separating the middle category. It is to be noted that also in trees, exploiting the order in categories means that implicitly *split variables* are fitted, that is, the variables considered in Chap. 4, which splits the response categories into two groups based on the order of categories. Whenever the trait of interest is modeled within a component of the tree a (conditional) split variable is used. If one wants to fully exploit the order of categories each component that refers to a split variable should include the trait of interest in addition to other parameters like response style parameters.

Hierarchical structured models but with a focus on regression, not item response theory, have been considered by Tutz (1989), Peyhardi et al. (2016), Tutz (2020a). While many authors considered building blocks with a discrimination parameter De Boeck and Partchev (2012) use a generalized linear model type framework without discrimination parameters but consider also latent-variable trees.

9.5 A Taxonomy of Models Including Tree-Based Models

A taxonomy of classical polytomous response models has already been given in Chap. 4 (Fig. 4.21). Figure 9.9 shows an extended hierarchy that includes tree-based models. The only conditional models in the hierarchy considered earlier are the adja-

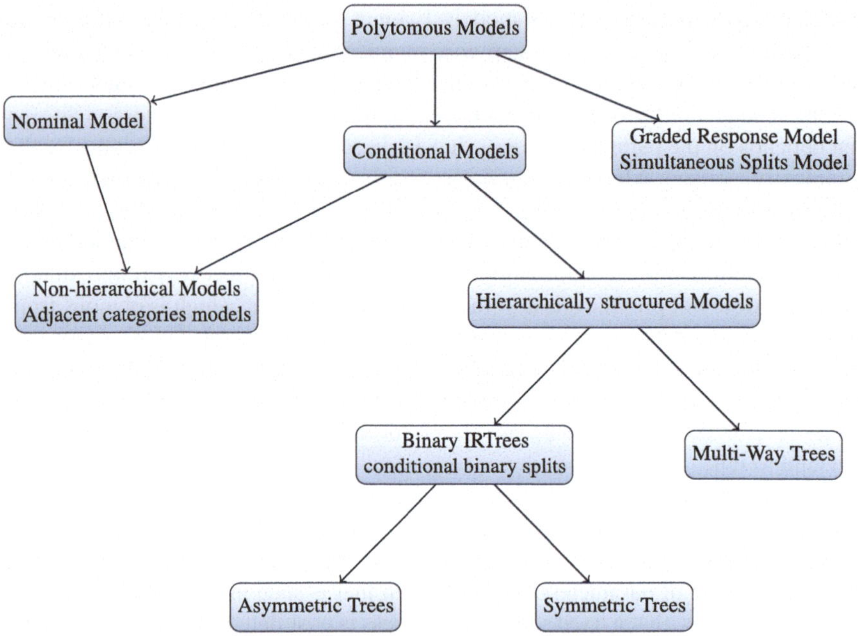

Fig. 9.9 Hierarchy of polytomous models

cent categories model and the sequential model. As has been shown in the present chapter, the sequential model is a specific model within the larger model class of tree-based or hierarchically structured models. Therefore, in the hierarchy shown in Fig. 9.9, the sequential model is replaced by the larger class of hierarchically structured models. The adjacent categories model is now referred to as non-hierarchical model to clarify the distinction to hierarchical models. There are two groups of conditional models. In the first group pairs of categories are compared by utilizing a binary response model to obtain, for example, the partial credit model and its simplified or extended versions. The second group is formed by hierarchical models. The crucial difference between non-hierarchical and hierarchical models is that in the former the conditions under which binary models are assumed to hold are overlapping. For example, in the partial credit model one binary submodel conditions on the categories $\{0, 1\}$, another submodel conditions on $\{1, 2\}$. Both conditions contain the category 1. This overlapping prevents a representation as a hierarchical model.

Hierarchically structured models can be divided into two types of models, binary IRTrees and multi-way trees. The former use only binary models to describe the conditional response in subsets of categories while the latter use traditional models with more than two categories as building blocks.

One can further distinguish between symmetric and asymmetric models. Symmetric models are in particular useful for Likert items to account for the symmetry in answer categories. Symmetric trees can be defined by considering subsets of categories $S_1, S_2 \subset \{1, \ldots, a\}$, where $a = m/2$ if m is even, and $a = (m-1)/2$ if m is odd. A tree is symmetric if for any (conditional) split between S_1 and S_2 there is a split between $S_3 = \{r|m-r+1, r \in S_2\}$ and $S_4 = \{r|m-r+1, r \in S_1\}$. An example with an even number of categories is the splitting structure shown in Fig. 9.2. If the number of categories is odd, and Likert items are considered, the first split typically distinguishes between the neutral category and the other categories. Although the visual appearance of corresponding trees shows some asymmetry, they are symmetric, and treat categories in a proper way, if the conditions of the definition are fulfilled. A classical example of a tree model that is not symmetric is the sequential model. It is, in particular, not invariant under the reverse permutation of categories; if the order of categories is reversed the corresponding sequential model differs from the sequential model for the original categories. In contrast, most symmetric models in common use are invariant under the reverse permutation, namely symmetric models, in which the response function $F(.)$ is a symmetric distribution function. The distinction between asymmetric symmetric models can be made for all hierarchically structured models. It is included in the taxonomy only for binary IRTrees, which have been investigated in the literature more intensively than other hierarchical models.

In general, the graded response model and the adjacent categories models can be used for any form of graded responses, in achievement tests as well as in the investigation of attitudes. Hierarchically structured models are somewhat different, and they are process models tailored to model a specific process. The sequential model assumes that levels of performance are reached successively, and therefore is most useful in items that are constructed with categories that represent successive solutions levels. Binary IRTrees and hierarchical partitioning approaches assume a specific conditional structure that aims at modeling the way how respondents generate a response. In hierarchically structured models, as in all conditional models, some care is needed when interpreting parameters since they typically are defined for pseudo-items. If each pseudo-item has its own person parameter interpretation refers to the conditional response captured in this pseudo-item. A more general interpretation applies only if the same trait is present in several pseudo-items.

9.6 Fitting Trees

Software that fits the variety of possible trees is not easy to obtain. An available package is *irtrees* De Boeck and Partchev (2012), which fits tree-based item response models of the generalized mixed model family, in particular binary trees. We consider a simple example using code from De Boeck and Partchev (2012). It also serves as an example for the difficulties that might arise when using trees as modeling tools.

Example 9.1 Verbal aggression
As an illustrating example we consider the verbal aggression data VerbAgg3, which is included in the package *irtrees*. The data are responses of 316 individuals to a questionnaire on verbal aggression inventory with 24 items. The three response

categories are "no", "perhaps", and "yes". De Boeck and Partchev (2012) fitted the tree shown in Fig. 9.10, which represents the sequential model. It is given in their notation with nodes represented by the binary variables Y_1^*, Y_2^*. They fitted a one-dimensional model, in which both nodes share a common latent trait, and a two-dimensional model, in which each node has its own latent trait by using: □

```
library("irtrees")
mapping <- cbind(c(0, 1, 1), c(NA, 0, 1))
VerbAgg3T <- dendrify(VerbAgg3[ , -(1:2)], mapping)
#  One-dimensional:
onedim <- glmer(value ~ 0 + item:node + (1 | person),
                          family = binomial, data = VerbAgg3T)
#Two-dimensional:
twodim <- glmer(value ~ 0 + item:node + (0 + node | person),
                         family = binomial, data = VerbAgg3T)
```

The comparison of the goodness-of-fit of the one- and two-dimensional model is seen as a test if the categories "no", "perhaps", and "yes" are ordered. De Boeck and Partchev (2012) consider it as strong evidence against a quantitative ordering of the response categories that the correlation between the two latent traits in the two-dimensional model is rather small (0.392), see Table 9.1, which shows the AIC values and the correlation between latent traits. Although it may be taken as a hint that categories are not used in an ordered way it can not be considered a general test of the ordinal hypothesis. Rather it is investigated if two-dimensions are needed given the sequential model in Fig. 9.10 holds. However, the sequential model, even when two latent scales are at play, is a model for ordered responses. It specifies the transition from category 0 to category 1 and then, given category 1 is reached, the transition from category 1 to category 2. It clearly uses the ordering and is not a model for nominal responses, and therefore can hardly be used to investigate if the ordering of categories is used by respondents.

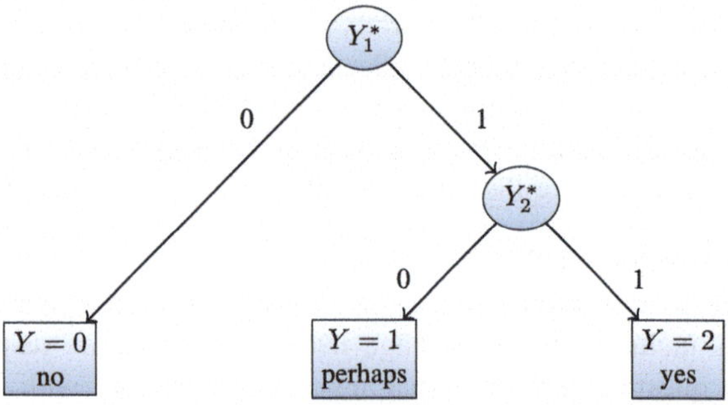

Fig. 9.10 Sequential model tree for three categories

Table 9.1 AIC values and correlations between latent traits for tree models (aggression data)

	Model	Sequential	Alternative	Non-ordered model sequential model
	One-dimensional	12778.23	8176.50	13671.016
	Two-dimensional	12377.99	8180.50	12558.724
	Correlation	0.39	0.10	−0.30

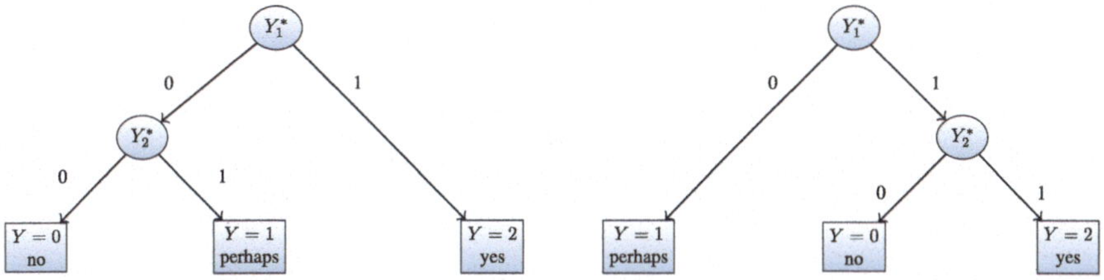

Fig. 9.11 Alternative sequential model tree (left) and non-ordered model tree (right)

In addition, the sequential model makes the strong assumption of a specific sequential decision process without strong evidence. There are alternative sequential mechanisms that could drive the response. Figure 9.11 shows two alternative trees that are based on other pseudo-items or mental queries. The tree on the left assumes that the first decision is between category groups {0, 1} and {2} and then conditionally between categories 0 and 1. It is referred to as "Altèrnative sequential model". The tree on the right assumes that the category "perhaps" is considered a separate category. The first decision is between "perhaps" and a clear statement "yes" or "no". Then it is distinguished between "yes" and no conditionally. This latter tree does not use the ordering of categories and is referred to as "Non-ordered model". As is seen from Table 9.1, the alternative sequential model shows much better fit than the sequential model. The one-dimensional alternative sequential model has the smallest AIC value among all the models considered. Although that makes it the best fitting and also parsimonious model it cannot be concluded that the underlying response model is the alternative sequential model. Goodness-of-fit cannot be considered a tool to decide which driving mechanism is the true one.

It should be noted that the one-dimensional non-ordered model is not a sensible model. Assuming that there is one latent trait for these choices is counter intuitive. It is only included for completeness. The model with two latent traits makes sense and even yields a smaller AIC value than the one-dimensional sequential model. The negative correlation between latent traits indicates that a tendency to prefer "yes" over "no" comes with a tendency to prefer "perhaps" over a distinct positioning, "yes" or "no".

Chapter 10
Differential Item Functioning

Differential item functioning (DIF) is a well known problem in item response theory. It occurs if the probability of a correct response among equally able persons differs in subgroups, for example, if the difficulty of an item depends on the membership to a racial, ethnic or gender subgroup. Then the performance of a group can be lower because these items are related to specific knowledge that is less present in this group. The effect is measurement bias and possibly discrimination. Various forms of differential item functioning have been considered in the literature, see, for example, Holland and H. (1993); Millsap and Everson (1993); Zumbo (1999); Osterlind and Everson (2009); Rogers (2005). Magis et al. (2010) gave an excellent overview of the existing dichotomous differential item functioning methods.

There are two basic approaches to DIF detection methods: those that use statistical methods for categorical data without assuming a specific IRT model to hold and those that aim at detecting DIF items given a specific item response model. The former are referred to as non-IRT methods, the latter as IRT methods, some authors use the terms parametric and nonparametric methods instead.

Most methods investigate DIF that is associated to group membership, in the simplest case there are just two groups, a focal group and a reference group, for example men and women. Then it is investigated if the performance of groups differ for specific items given varying ability levels. In the following we will consider some basic methods for DIF induced by group membership (non-IRT methods and IRT methods) but will in particular consider more recently developed methods that link DIF to a vector of person-specific covariates.

One more distinction of methods is useful: *uniform* DIF is present if the (scaled) differences in the probabilities of solving an item of subjects from different groups but with the same ability level do not depend on the common ability level. In *non-uniform* DIF scenarios the differences are not constant across ability levels and crossing item response curves may occur.

G. Tutz, *A Short Guide to Item Response Theory Models*, Statistics for Social and Behavioral Sciences, https://doi.org/10.1007/978-3-031-87271-6_10

10.1 Non-IRT Methods

The traditional approach to identify items that carry DIF is based on test statistics. For each item a test is performed that shows if the item has different difficulties in subgroups that have to be defined by the experimenter. Test statistics have been proposed by Thissen et al. (1993), **?**, Holland and Thayer (1988), Kim et al. (1995), and Raju (1988). Mixed model approaches were proposed by Van den Noortgate and De Boeck (2005) and Bayesian approaches have been developed by Soares et al. (2009).

Mantel-Haenszel Tests

A popular method for DIF detection is the MH method. It uses the Mantel-Haenszel test statistic (Mantel and Haenszel, 1959) and has been propagated by Holland and Thayer (1988), Verguts and De Boeck (2001). It tests whether there is an association between group membership and item response, conditionally upon the total test score.

Typically there are just two groups, a reference group and a focal group, and the responses are binary $Y_{pi} \in \{0, 1\}$. Given this setting let $n^{(s)}$ denote the number of examinees with sum score s for a fixed item. The $n^{(s)}$ examinees are cross-classified into a (2×2) contingency table with group membership and type of response as variables.

	item response 1	item response 0	
reference group	$n_{r1}^{(s)}$	$n_{r0}^{(s)}$	$n_{r+}^{(s)}$
focal group	$n_{f1}^{(s)}$	$n_{f0}^{(s)}$	$n_{f+}^{(s)}$
	$n_{+1}^{(s)}$	$n_{+0}^{(s)}$	$n^{(s)}$

The four cell counts of the table are given by

$n_{rj}^{(s)}$, number of responses with value j in the reference group, $j \in \{0, 1\}$

$n_{fj}^{(s)}$, number of responses with value j in the focal group, $j \in \{0, 1\}$

The marginal sums are $n_{r+}^{(s)} = n_{r1}^{(s)} + n_{r0}^{(s)}$, $n_{f+}^{(s)} = n_{f1}^{(s)} + n_{f0}^{(s)}$, $n_{+1}^{(s)} = n_{r1}^{(s)} + n_{f1}^{(s)}$, $n_{+0}^{(s)} = n_{r0}^{(s)} + n_{f0}^{(s)}$.

The MH test is a non-model-based test for conditional independence in a collection of (2×2) tables (for each observed sum score there is one table). The statistic has the form

$$\mathrm{MH} = \frac{(\sum_s |n_{r1}^{(s)} - \hat{\mathrm{E}}(n_{r1}^{(s)})| - 0.5)^2}{\hat{\mathrm{var}}(n_{r1}^{(s)})},$$

where $\hat{\mathrm{E}}((n_{r1}^{(s)})) = n_{r+}^{(s)} n_{+1}^{(s)} / n^{(s)}$, $\hat{\mathrm{var}}(n_{r1}^{(s)}) = n_{r+}^{(s)} n_{f+}^{(s)} n_{+1}^{(s)} n_{+0}^{(s)} / ((n^{(s)})^2 (n^{(s)} - 1))$. Under the null hypothesis of no conditional association between item response and group membership the MH statistic follows asymptotically a chi-squared distribution with one degree of freedom. An item is considered to carry DIF if the MH statistic value is larger than a critical value of the asymptotic chi-squared distribution. If the number of groups is larger than two the generalized Mantel-Haenszel test can be used, which is given, for example, in Agresti (2013), Sect. 8.4.3.

An associated summary measure of conditional association is given by

$$\hat{\theta}_{\text{MH}} = \frac{\sum_s n_{r1}^{(s)} n_{f0}^{(s)} / n^{(s)}}{\sum_s n_{f1}^{(s)} n_{s0}^{(s)} / n^{(s)}}.$$

The logarithm $\log(\hat{\theta}_{\text{MH}})$ can also be used to test DIF. As a measure of conditional association the size can be used to indicate the strength of DIF. If $\log(\hat{\theta}_{\text{MH}})$ is far from zero one might suspect DIF. According to Holland and Thayer (1985) an item shows large DIF if $|2.35 \log(\hat{\theta}_{\text{MH}})| > 1.5$, and moderate DIF if the value is between 1 and 1.5; otherwise, it is neglected. The method has also been called ETS Delta scale method.

The basic concept of this type of DIF detection is to use the sum score as a proxy for the ability of persons. If there is association between the item response (correct or not) and the group membership when accounting for the ability (through the sum score) this is seen as an indicator of DIF. It is implicitly assumed that the sum score is a valid indicator of the ability.

Logistic Regression: Comparison of Groups

A test score based method to detect uniform DIF was proposed by Swaminathan and Rogers (1990) and has been extended to multiple groups by Magis et al. (2011). The basic concept is to fit a logistic model for answering an item correctly given the test score and the group membership. The model for examining item i has the form

$$\log\left(\frac{P(Y_{pi} = 1|S_p, M_p)}{P(Y_{pi} = 0|S_p, M_p)}\right) = \eta_{pi} = \beta_{0i} + S_p\beta_i + M_p\gamma_{ig}, \quad p = 1, \ldots, P, \tag{10.1}$$

where S_p is the test score of person p and M_p is the group membership indicator with $M_p = 1$ if person p is from group g and $M_p = 0$ otherwise. If there are G groups, one of the parameters $\gamma_{i1}, \ldots, \gamma_{iG}$ has to be set to zero; for example, by setting $\gamma_{i1} = 0$ the first group is implicitly chosen as the reference group. In this model the parameter β_{0i} is the counterpart of the item difficulty and the parameter β_i is the counterpart of an discrimination parameter. Again the test scores are considered as proxies for the abilities of persons.

For the detection of DIF the most interesting parameters are the group-specific parameters $\gamma_{i1}, \ldots, \gamma_{iG}$. They represent the differential item functioning. In the simplest case of two groups, a reference group and a focal group, one chooses $\gamma_{i1} = 0$ for the reference group and only one parameter determines DIF. For example, with groups defined by gender with female as the reference group the predictors are

$$\tilde{\beta}_{0i} + \gamma_{i,male} \text{ for males and } \tilde{\beta}_{0i} \text{ for females,}$$

where $\tilde{\beta}_{0i} = \beta_{0i} + S_p\beta_i$. If $\gamma_{i,male} \neq 0$ one has DIF in item i generated by gender. If, for example, $\gamma_{i,male} > 0$ the probability of solving the item is for men enlarged whatever the ability of the person captured in $\tilde{\beta}_{0i}$.

DIF detection within the logistic regression framework typically uses likelihood ratio statistics that test the null hypothesis $H_0 : \gamma_{i1} = \cdots = \gamma_{iG} = 0$. If the hypothesis is rejected item i is considered as a DIF item. Each item is tested separately at significance level α with the degrees of freedom equal to $G - 1$, depending on the number of groups.

Logistic Regression: DIF Generated by a Vector of Covariates

The tests considered so far are limited to the consideration of subgroups. That means if one suspects item difficulties to depend on age one has to know the age groups before testing. Thus age has to be split into two or more intervals without knowing which ones are relevant. Moreover, the approaches do not allow to investigate DIF generated by more than one variable. However, the basic concept can be simply extended to include continuous (and categorical) variables that might induce DIF. Let $\boldsymbol{x}_p^\top = (x_{p1}, \ldots, x_{pm})$ be a vector of person-specific explanatory variables of length m. Then, model (10.1) can be extended to

$$\log\left(\frac{P(Y_{pi} = 1|S_p, \boldsymbol{x}_p)}{P(Y_{pi} = 0|S_p, \boldsymbol{x}_p)}\right) = \eta_{pi} = \beta_{0i} + S_p\beta_i + \boldsymbol{x}_p^\top\gamma_i. \tag{10.2}$$

The new parameters in model (10.2) are the weights γ_i contained in the term $\boldsymbol{x}_p^\top\gamma_i$. The comparison of multiple groups is just a special case. Setting the first group as reference one defines the vector of explanatory variables $\boldsymbol{x}_p^\top = (x_{p2}, \ldots, x_{pG})$, where $x_{pg} = 1$ if person p is from group g and 0 otherwise. Then, the vector of parameters for item i is $\gamma_i^\top = (\gamma_{i2}, \ldots, \gamma_{iG})$. Uniform DIF is present in this item if $\gamma_i \neq \boldsymbol{0}$. To investigate DIF one uses a global test for the whole parameter vector, $H_0 : \gamma_i = \boldsymbol{0}$. The alternative hypothesis is that at least one of the parameters is unequal to zero. The hypotheses are tested separately for each item at significance level α. Due to the design of the test the approach identifies the items that carry DIF but does not contain any information about the components of $\boldsymbol{x}_p$ that are responsible for DIF. However, if DIF is detected for some items one can investigate which variables are responsible by testing the components of the weights one at a time by testing the null hypotheses $H_0 : \gamma_{ij} = \boldsymbol{0}$, $j = 1, \ldots, m$ separately.

Logistic Regression for Non-uniform DIF

The approaches considered so far aimed at detecting uniform DIF, which is present if the differences in the probabilities of solving an item of subjects from different groups but with the same ability level do not depend on the common ability level. An extended form of the logistic model can be used to address non-uniform DIF.

Let the population be divided into two groups, and M_p denote the indicator of group membership, $M_p = 1$ for the focal group, and $M_p = 0$ for the reference group. The model

$$\log\left(\frac{P(Y_{pi}=1|S_p,g)}{P(Y_{pi}=0|S_p,g)}\right)=\eta_{pi}=\beta_{0i}+S_p\beta_i+M_p\gamma_i+S_pM_p\alpha_i,\quad p=1,\ldots,P, \tag{10.3}$$

contains the additional interaction term $S_pM_p\alpha_i$ between sum score and group membership. Tests as the likelihood ratio test, the Wald test or the score test can be used to examine if parameters are needed. Rejection of the null hypothesis $\alpha_i=0$ indicates that non-uniform DIF is present. Then, the corresponding predictors are

$$\tilde{\beta}_{0i}+\gamma_i+S_p\alpha_i \text{ for the focal group and } \tilde{\beta}_{0i} \text{ for the reference group,}$$

where $\tilde{\beta}_{0i}=\beta_{0i}+S_p\beta_i$. That means that the linear predictors in the focal group is shifted by $\gamma_i+S_p\alpha_i$. It represents non-uniform DIF since the shifting varies with S_p, which is considered a proxy for the ability. Thus, DIF depends on the ability.

If the null hypothesis $\alpha_i=0$ is not rejected, still uniform DIF may be present. To exclude DIF one has to examine the null hypothesis $\gamma_i=0$. If it is also not rejected one might assume that the item is DIF-free.

An extended version of the model can accommodate for multiple groups by using a multiple group indicator M_p and the term $M_p\gamma_{ig}+S_pM_p\alpha_{ig}$instead of $M_p\gamma_i+S_pM_p\alpha_i$. Also a general vector of person-specific covariates $\boldsymbol{x}_p$ can be used in the interaction term yielding the model

$$\log\left(\frac{P(Y_{pi}=1|S_p,g)}{P(Y_{pi}=0|S_p,g)}\right)=\eta_{pi}=\beta_{0i}+S_p\beta_i+\boldsymbol{x}_p^T\boldsymbol{\gamma}_i+S_p\boldsymbol{x}_p^T\boldsymbol{\alpha}_i,\quad p=1,\ldots,P. \tag{10.4}$$

The interaction $S_p\boldsymbol{x}_p^T\boldsymbol{\alpha}_i$ allows to investigate if the variables contained in $\boldsymbol{x}_p$ induce non-uniform DIF.

In the literature different strategies were proposed how to test for the significance of DIF generated by groups, see, for example, Zumbo (1999) and Magis et al. (2011). Similar strategies can be used when testing for DIF in the extended logistic regression model (10.4). The first strategy is to test for both types of DIF effects simultaneously by examining the null hypothesis $H_0:\boldsymbol{\gamma}_i=\boldsymbol{\alpha}_i=0$ given model (10.4). That means DIF is investigated by using a global test for the whole parameter vector $(\boldsymbol{\gamma}_i,\boldsymbol{\alpha}_i)$. DIF is considered as being present (in any form) if the test rejects the null hypothesis, meaning that at least one of the parameters $\gamma_{ij},\ \alpha_{ij},\ j=1,\ldots,m$, differs from zero.

A second strategy is to explicitly test for non-uniform DIF. Using the extended logistic model (10.4) one investigates the null hypothesis $H_0:\boldsymbol{\alpha}_i=\boldsymbol{0}$ for each item. Non-uniform DIF is considered as being present if the hypothesis is rejected, meaning that at least one parameter α_{ij} differs from zero. An illustration of the method is given later in Example 10.5 together with tree-based methods.

Logistic Regression: Problems and Alternative Methods

The logistic regression approach to DIF detection is not without problems. One problem is that the test scores are used as proxys for the abilities of persons. However, test scores as the number of solved items are sufficient statistics for ability parameters only if the Rasch model holds, and if no DIF is present. By including parameters that represent discrimination parameters the regression models do not correspond to Rasch models but are more in line with the 2PL model. Another problem is that when testing it is assumed that all other items are free of DIF, which is an assumption that typically does not hold. Also, there is a multiple testing problem since the regression models are fitted separately for each item.

These problems are addressed by the *lasso DIF* method proposed by Magis et al. (2015). They consider the model

$$\log\left(\frac{P(Y_{pi}=1|S_p,M_p)}{P(Y_{pi}=0|S_p,M_p)}\right)=\eta_{pi}=\beta_{0i}+S_p\beta+M_p\gamma_{ig},\quad p=1,\dots,P,$$

where M_p is a group indicator as in model (10.1). In addition, the weight on the test score, β, does not depend on the item. Thus, there are no discrimination parameters that vary across items. The model does not investigate DIF separately for items but all item parameters are estimated simultaneously. Instead of using simple maximum likelihood (ML) estimation Magis et al. (2015) used penalized ML estimation. With all item parameters collected in the parameter vector $\boldsymbol{\beta}$, they maximize the lasso-type penalized log-likelihood

$$l_{\text{pen}}(\boldsymbol{\beta})=l(\boldsymbol{\beta})-\lambda\sum_{i=1}^{I}|\gamma_{ig}|,$$

where $l(.)$ is the common log-likelihood of the model and λ is a tuning parameter. If λ is very small the estimates are close to the ML estimate, if it is large the estimates of the parameters γ_{ig} become small. The advantage of the lasso penalty $\sum_i|\gamma_{ig}|$, which was originally proposed by Tibshirani (1996), is that it enforces selection of parameters since for appropriately chosen λ part of the parameters are estimated as being zero while the rest of the parameters are shrunk but still unequal zero. For details, how to select λ see Magis et al. (2015).

The concept of penalized estimation should also be useful when investigating DIF induced by a vector of explanatory variables. Lasso type penalties should be able to select the variables that generate DIF and the items that carry DIF simultaneously. For IRT models such methods are already available (see next section).

10.2 IRT Methods

Several methods have been proposed that are able to detect DIF that is associated with group membership, they include likelihood ratio test methods (Thissen et al., 2013), Lord's chi-square test (**?**), and the Raju method (Raju, 1988). In the following these methods are not considered, instead we focus on more general methods that are able to detect DIF generated by more than one variable. They can also be used if only one variable or a group indicator is to be investigated.

DIF in Rasch Models Using Regularized Estimates

A method that is able to examine DIF that is associated with a vector of explanatory variables and selects which variables are responsible for DIF in single items has been proposed by Tutz and Schauberger (2015).

Let us consider the binary Rasch model $P(Y_{pi} = 1) = \exp(\theta_p - \delta_i)/(1 + \exp(\theta_p - \delta_i))$. In the DIF Rasch model, which allows the item parameters to depend on covariates that characterize the person, the item parameter is replaced by a linear form that includes the vector of person-specific explanatory variables $\boldsymbol{x}_p$. With item-specific parameter $\boldsymbol{\gamma}_i$ one considers the model

$$\log\left(\frac{P(X_{pi} = 1)}{P(X_{pi} = 0)}\right) = \theta_p - (\beta_i + \boldsymbol{x}_p^T \boldsymbol{\gamma}_i) \tag{10.5}$$

The essential advantage of model (10.5) is that the item-specific parameter refers to all the candidate variables that are under suspicion to induce differential item functioning, for example, age, race, gender, and other variables. If one component in the vector $\boldsymbol{\gamma}_i$ is unequal to zero the item is group-specific, and the respective parameter indicates which of the variables is responsible for the differential item functioning.

The challenge of the model is how to estimate the large number of parameters and to determine which parameters have to be considered as unequal zero. The basic assumption is that most of the parameters do not depend on the group, but some can. One wants to detect these items and know which ones of the explanatory variables induce DIF.

To this end maximum likelihood (ML) estimation is replaced by penalized maximum likelihhood. Let $\boldsymbol{\alpha}$ denote the total vector of parameters, $\boldsymbol{\alpha}^T = (\boldsymbol{\theta}^T, \boldsymbol{\beta}^T, \boldsymbol{\gamma}_1^T, \ldots, \boldsymbol{\gamma}_I^T)$. Then one maximizes the penalized log-likelihood

$$l_{\text{pen}}(\boldsymbol{\alpha}) = l(\boldsymbol{\alpha}) - \lambda J(\boldsymbol{\alpha}),$$

where $l(.)$ is the common log-likelihood of the model and $J(\boldsymbol{\alpha})$ is a penalty term that penalizes specific structures in the parameter vector. The parameter λ is a tuning parameter that specifies how seriously the penalty term has to be taken. A widely used penalty term in regression problems is $J(\boldsymbol{\alpha}) = \boldsymbol{\alpha}^T \boldsymbol{\alpha}$, that is, the squared length of the parameter vector. The resulting estimator is known under the name ridge estimator, see Hoerl and Kennard (1970) for linear models and Nyquist (1991), Segerstedt (1992) for generalized linear models. If $\lambda = 0$ maximization yields the ML estimate. If $\lambda > 0$ one obtains parameters that are shrunk toward zero. In the extreme case $\lambda \to \infty$ all parameters are set to zero. The ridge estimator with small $\lambda > 0$ stabilizes estimates but does not select parameters, which is the main objective here.

A much more useful penalty that enforces selection of parameters that are unequal zero is the L_1-penalty or lasso (for least absolute shrinkage and selection operator) (Tibshirani, 1996). Let us first consider the simple case of a *univariate* explanatory variable, which, for example, codes gender. Then the *lasso penalty for differential item functioning* (DIF-lasso) is given by

$$J(\boldsymbol{\theta}^T, \boldsymbol{\beta}^T, \gamma_1^T, \ldots, \gamma_I^T) = \sum_{i=1}^{I} |\gamma_i|.$$

The penalty term contains only the parameters that are responsible for differential item functioning, therefore only the parameters that carry information on DIF are penalized. Again, if $\lambda = 0$ maximization yields the full ML estimate. For very large λ all the γ-parameters are set to zero. In the extreme case $\lambda \to \infty$ the Rasch model is fitted without allowing for differential item functioning. The interesting case is in between, when λ is finite and $\lambda > 0$. Then, for fixed λ, some of the parameters are set to zero while others take values unequal zero. With a carefully chosen tuning parameter λ the parameters that yield estimates $\hat{\gamma}_i > 0$ are the ones that show DIF.

In the general case with a vector of covariates that potentially induce DIF a more appropriate penalty is a modification of the grouped lasso (Yuan and Lin, 2006; Meier et al., 2008). Let $\boldsymbol{\gamma}_i^T = (\gamma_{i1}, \ldots, \gamma_{im})$ denote the vector of modifying parameters of item i, where m denotes the length of the person-specific covariates. Then the *group lasso penalty for differential item functioning* (DIF-lasso) is

$$J(\boldsymbol{\theta}^T, \boldsymbol{\beta}^T, \boldsymbol{\gamma}_1^T, \ldots, \boldsymbol{\gamma}_I^T) = \sum_{i=1}^{I} \|\boldsymbol{\gamma}_i\|,$$

where $\|\boldsymbol{\gamma}_i\| = (\gamma_{i1}^2 + \cdots + \gamma_{im}^2)^{1/2}$ is the L_2-norm of the parameters of the ith item with m denoting the length of the covariate vector. The penalty encourages sparsity in the sense that either $\hat{\boldsymbol{\gamma}}_i = \mathbf{0}$ or $\gamma_{is} \neq 0$ for $s = 1, \ldots, m$. Thus the whole group of parameters collected in $\boldsymbol{\gamma}_i$ is shrunk simultaneously toward zero. For a geometrical interpretation of the penalty, see Yuan and Lin (2006). The effect is that in a typical application only some of the parameters get estimates $\hat{\boldsymbol{\gamma}}_i \neq \mathbf{0}$. These correspond to items that show DIF.

An important issue in penalized estimation is the choice of the tuning parameter λ. A criterion that yields a satisfying balancing is the BIC (Schwarz, 1978) with the degrees of freedom for the group lasso penalty determined by a method proposed by Yuan and Lin (2006). For the details of the adaptation to DIF modeling see Tutz and Schauberger (2015). More recently, Gürer and Draxler (2023) proposed a lasso approach to DIF in Rasch models that exploits the advantages of conditional likelihood estimation, which works also in Rasch models that contain covariates, extensions of the model have been proposed by Draxler et al. (2024), Draxler and Kurz (2025). Chen et al. (2023) proposed a method based on marginal estimates that allows different groups to have different latent trait distributions.

Example 10.1 Knowledge Data

We use data from an online quiz for testing one's general knowledge conducted by the weekly German news magazine SPIEGEL. The data have also been used by Strobl et al. (2015) and Tutz and Schauberger (2015). The presentation follows Tutz and Schauberger (2015). The 45 test questions were from five topics, politics, history, economy, culture, and natural sciences. We use the same sub sample as Strobl et al. (2015) consisting of 1075 university students from Bavaria, who had all been assigned a particular set of questions. The covariates that we included as potentially inducing DIF are gender, age, semester of university enrollment, an indicator for whether the student's university received elite status by the German excellence initiative (elite), and the frequency of accessing SPIEGEL's online magazine (spon). Figure 10.1 shows as an example the coefficient build-ups for the covariate gender. It shows the estimates of γ_i as a function of the tuning parameter. If $\lambda \rightarrow \infty$ all parameters are set to zero. With growing λ they take values unequal zero. Instead of using λ itself on the x-axis the size of the estimated parameters is used ($||\gamma_i||/\max ||\gamma_i||$). At the path point that was selected by the BIC criterion (dashed vertical line), 16 of the 45 items showed DIF, which is not surprising because it is not a carefully constructed test that really focusses on one latent dimension. In Fig. 10.2, the estimated effects of the items containing DIF are visualized. The upper panel shows the profile plots of the parameters for the included covariates. For each item with DIF one profile is given. The lower panel shows the strengths of the effects in terms of the absolute value of the coefficients. One boxplot refers to the absolute values of the 16 parameters for one covariate. It is seen that the strongest effects are found for the covariate gender, the weakest effects are in the variable elite, which measures the status of the university where the student is enrolled. It should be noted that the importance of the single covariates for the DIF can be measured by the absolute values of their coefficients since all covariates were standardized.

In Fig. 10.2 (upper panel) four items are represented by dashed lines. They showed the strongest DIF in terms of the L_2-norm of the estimated parameter vector. All of them refer to economics. For illustration, these four items are considered in more detail. They are

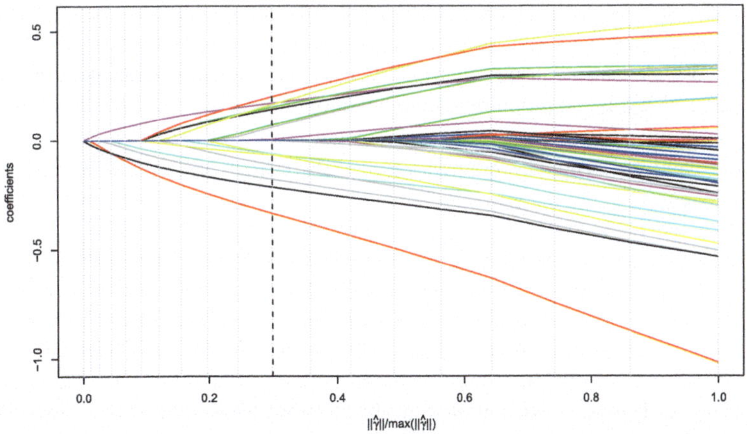

Fig. 10.1 Coefficient build-ups for covariate gender in Quiz Data; dashed vertical line indicates BIC-optimal path point. Reprinted with permission from Tutz and Schauberger (2015)

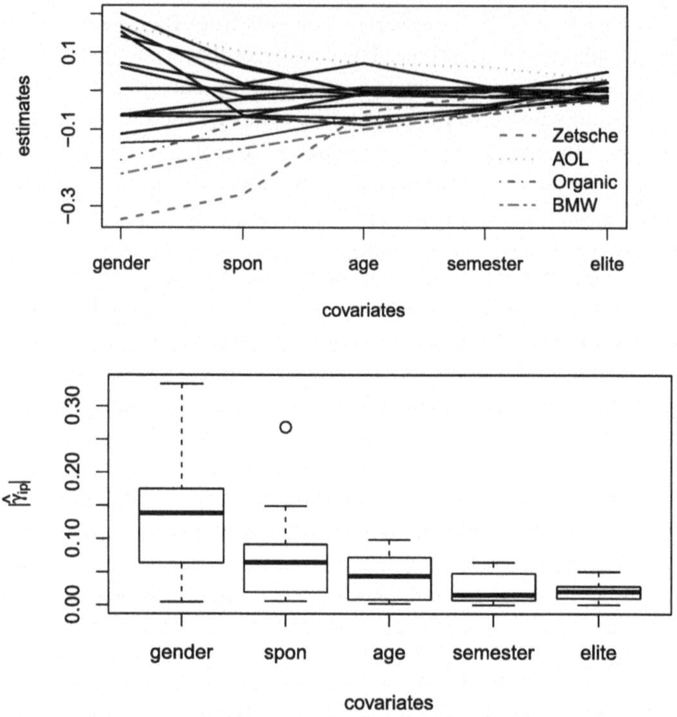

Fig. 10.2 Upper panel: profile plot for coefficient estimates of items with DIF, profiles of the four items with highest DIF are highlighted; lower panel: boxplots of absolute values of coefficient-estimates for items with DIF. Reprinted with permission from Tutz and Schauberger (2015)

- Zetsche: "Who is this?" (a picture of Dieter Zetsche, the CEO of the Daimler AG, maker of Mercedes cars, is shown).
- AOL: "Which internet company took over the media group Time Warner?"
- Organic: "What is the meaning of the hexagonal 'organic' logo?" (Synthetic pesticides are prohibited)
- BMW: "Which German company took over the British automobile manufacturers Rolls-Royce?"

The profiles for the items Zetsche, AOL, and BMW are quite similar. They are distinctly easier for male participants and for frequent visitors of SPIEGELonline. The item Organic shows a quite different shape being definitely easier for females. It is also easier to solve for students that are not frequent visitors of SPIEGELonline. The item differs from the other three items because it refers more to a broad education than to current issues. Also, females might be more interested in (healthy) food in general. In this respect female students and students that do not follow the latest news seem to find the item easier. Therefore the different profiles.

Example 10.1 is based on penalized Ml estimation as propagated by Tutz and Schauberger (2015) and was fitted by using the R package *DIFlasso*. Alternative versions that use penalized marginal ML methods were given by Schauberger and Mair (2020). The latter also allows to detect differential item functioning in generalized partial credit models and can be fitted by using the R package *GPCMlasso* (Schauberger, 2024a). An alternative way to obtain regularized estimates is boosting, which for DIF detection in Rasch models has been investigated by Schauberger and Tutz (2016).

DIF in Thresholds Models

It is straightforward to model DIF in thresholds models, which comprise various classical models as the graded response model. The basic threshold model has the form $P(Y_{pi} > y) = F(\alpha_i(\theta_p - \delta_i(y)))$ (Chap. 6). If one suspects that covariates may modify the response behaviour it can be tested by including them explicitly in the explanatory term. Let $\boldsymbol{x}_p$ be a person-specific vector of covariates. Then, the person parameter θ_p is replaced by $\theta_p - \boldsymbol{x}_p^T\gamma_i$ yielding the DIF thresholds model

$$P(Y_{pi} > y|\theta_p, \alpha_i, \delta_i(.)) = F(\alpha_i(\theta_p - \boldsymbol{x}_p^T\gamma_i - \delta_i(y))), \tag{10.6}$$

where the parameter γ_i is item-specific and represents the effect on the response in item i. The extended model can be estimated by marginal log-likelihood.

Example 10.2 Fears Data

Let us consider the fears data from Example 6.2 with covariates gender (1: female; 0: male) and age in years. Table 10.1 shows the estimates for the basic model with-

Table 10.1 Parameter estimates for the fears data with logit difficulty function, Gumbel distribution response function without covariate and with covariates, the last two columns show the z-values of parameter estimates of covariate parameters

	Item	Parameters				z-values	
		Intercepts	Slopes	Age	Gender	z-Age	z-Gender
1		−0.751	1.0259				
2		−1.459	1.2609				
3		−2.134	1.3365				
4		−0.360	1.2061				
5		−1.125	1.1303				
	Log-lik	−1713.209					
1	Refugee	0.072	1.028	−0.011	−0.487	−2.138	−2.548
2	Climate change	−1.055	1.262	−0.002	−0.642	−0.326	−3.328
3	Terrorism	−0.954	1.393	0.017	−0.756	−3.379	−3.997
4	Globalization	0.416	1.201	−0.010	−0.491	−2.085	−2.603
5	Nuclear energy	−0.385	1.131	−0.013	−0.089	−2.632	−0.474
	Log-lik	−1692.576					

out covariates and the model with covariates age and gender (continuous response, $\alpha_i = 1$). It is seen that all items show significant covariate effects for at least one of the covariates (z-values given in the last two columns). With the exception of climate change, older respondents tend to be more afraid than younger respondents, females have for all items higher fear levels than males. The necessity of covariates is also supported by testing. The log-likelihood test that compares the model without covariates to the model with covariates is 41.266 on 10 df. Thus, differential item functioning is present.

10.3 Recursive Partitioning: Tree-Based Approaches

Recursive partitioning also known as tree-based modeling is an efficient nonparametric regression and classification method that is visualized by trees. The method is conceptually very simple. In regression problems there is a dependent variable Y and a vector of explanatory variables $\boldsymbol{x}^T = (x_1, \ldots, x_m)$ from a m-dimensional predictor space. A regression tree is obtained by successively splitting the predictor space by using just one variable to obtain a new partition. The starting point is the whole m-dimensional predictor space. One considers simple splits of the form $x_j \leq c, x_j > c$ in all covariates x_j and all sensible constants c. Then one selects the "best" split, that is, the split which best explains the variation of the dependent variable in the predictor space. The selected split yields a partition $A_1 = \{\boldsymbol{x}|x_{j_0} \leq c_0\}, A_2 = \{\boldsymbol{x}|x_{j_0} \leq c_0\}$, where x_{j_0} and c_0 denote the selected variable and constant, respectively. The first partition represents the root node. Then the predictor spaces A_1 and A_2 are considered, and the explanatory power of partitions of the form

$$A \cap \{x_j \leq c\}, \quad A \cap \{x_j > c\},$$

are examined, where $A \in \{A_1, A_2\}$. Again one selects the best partition with all variables and constants as candidates to obtain a new partition $A_n \cap \{x_{j_n} \leq c_n\}$, where A_n is the selected subspace, x_{j_n} is the selected variable, and c_n the selected constant. The new partition corresponds to a node in the tree. The procedure is repeated with A representing the subspace generated in previous nodes. The resulting tree describes a partitioning of the predictor space into rectangles. Various choices have to be made when growing a tree, in particular a criterion for the explanatory power of explanatory variables has to be chosen and a stop criterion is needed that determines when to stop growing the tree.

Recursive partitioning has its roots in automatic interaction detection (Morgan and Sonquist, 1963). The most popular modern version is due to Breiman et al. (1984) and is known by the name *classification and regression trees*, often abbreviated as CART. Alternative approaches are the C4.5 algorithm (Quinlan, 1986, 1993), or the recursive partioning framework based on conditional inference proposed by Hothorn et al. (2006). An overview with a focus on psychometrics was given by Strobl et al. (2009).

The mentioned literature and the basic procedure that has been described briefly refers to regression and classification trees. When using trees to detect DIF the method has to be adapted accordingly. It should be noted that the DIF trees considered in the following are conceptually different from the IRTrees considered in Chap. 9. IRTrees consider splits in the response categories whereas the trees considered here refer to partitions of the predictor space.

Rasch Trees

A method of DIF detection in Rasch models that uses recursive partitioning techniques has been proposed by Strobl et al. (2015) under the name Rasch trees. They propose to recursively partition the covariate space to identify regions of the covariate space in which DIF occurs. In the investigated regions a Rasch model that includes covariates is fitted. Regions are suspected to be relevant if the parameter estimates in the considered regions differ strongly. Therefore, regions in the covariate space are identified that show different difficulties. A defining feature of the method is that splits in the explanatory variables are identified that show the strongest differences in *all of the item difficulties simultaneously*. Therefore, it detects regions of the covariate space that are linked to DIF but does not detect the items that are responsible.

Example 10.3 Knowledge Data

Let us consider again the knowledge data from Example 10.1. To test for DIF five covariates were included: gender (0: female, 1: male), age, number of matriculated semester, elite status of the university (0: no, 1: yes) and the frequency of accessing Spiegel's online magazine (spon) from 1 (never) to 7 (daily). Rasch trees methodology yields a tree with splits in two variables, gender and spon shown in Fig. 10.3. These variables are found to induce DIF and one finds four groups that differ in terms of item difficulty. In each leaf of the corresponding tree the estimated difficulties are shown.

The fit is easily obtained by using the vignette *vignette("raschtree", package = "psychotree")*:

```
library(psychotree)
data(SPISA)
vignette("raschtree", package = "psychotree")
raschtree <- raschtree(spisa ~ age + gender +
            + semester + elite + spon, data = SPISA)
plot(raschtree)
```

As seen in Fig. 10.3 the Rasch trees method yields one tree for all of the items, which suggests that all items show DIF in the same covariates. It does not clearly identify the items that are responsible for the split and therefore for DIF. Consequently it is hard to identify those items that are affected by DIF and those that are not. There is no criterion provided to identify the DIF items. This is different in the item-focused approach considered in the following.

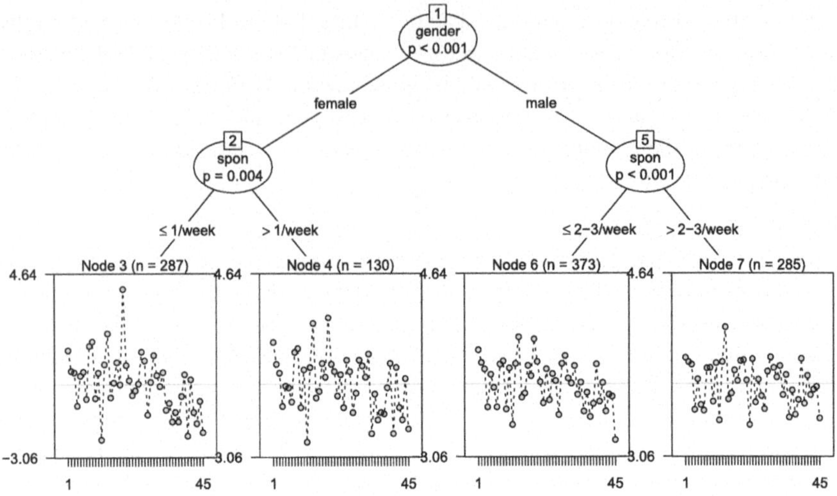

Fig. 10.3 Result of the analysis of the Spiegel online data by a Rasch tree

Item-Focused Trees for the Rasch Model

An alternative are item-focused trees (Tutz and Berger, 2016a), which focus on the detection of the items that carry DIF. Recursive partitioning is used on the item level not on the global level, which treats all items simultaneously, as in Rasch trees. The item-focused approach allows to detect the items that carry DIF but keeps the advantage that no pre-specified subgroups are needed.

Let $\boldsymbol{x}_p^T = (x_{p1}, \ldots, x_{pm})$ again denote a person-specific covariate vector. For the detection of DIF the first split means one examines for all the items, all the variables and possible splits of the corresponding variable the Rasch type model $P(Y_{pi} = 1) = F(\eta_{pi})$ with $F(.)$ denoting the logistic distribution function and predictor

$$\eta_{pi} = \theta_p - [\gamma_{il} I(x_{pj} \le c_j) + \gamma_{ir} I(x_{pj} > c_j)],$$

where $I(\cdot)$ denotes the indicator function with $I(a) = 1$ if a is true and $I(a) = 0$ otherwise. The model allows for DIF in item i by distinguishing just between two groups, $x_{pj} \le c_j$ and $x_{pj} > c_j$), which may be seen as focal and reference group constructed by a split of the jth variable at split-point c_j. The parameter γ_{il} denotes the item difficulty in the left node ($x_{pj} \le c_j$) and γ_{ir} the item difficulty in the right node ($x_{pj} > c_j$). One chooses that combination of item, variable and split that has the smallest p-value when tested for DIF, that is, in the examination of the null hypothesis $H_0 : \gamma_{il} - \gamma_{ir} = 0$. This selection yields the first split into left and right daughter nodes corresponding to the regions $x_{pj} \le c_j$ and $x_{pj} > c_j$.

Further splitting means that one of the subspaces, say $I(x_{pj} > c_j)$, is further split, for example, in variable s at cut point c_s, yielding the daughters

$$I(x_{pj} > c_j) I(x_{ps} \le c_s) \quad \text{and} \quad I(x_{pj} > c_j) I(x_{ps} > c_s).$$

and the linear predictor

$$\eta_{pi} = \theta_p - [\gamma_{il} I(x_{pj} \le c_j) + \gamma_{il}^{(n)} I(x_{pj} > c_j) I(x_{ps} \le c_s) + \gamma_{ir}^{(n)} I(x_{pj} > c_j) I(x_{ps} > c_s)],$$

where $\gamma_{il}^{(n)}, \gamma_{ir}^{(n)}$ are the weights on the new split. Then the item difficulty in the region $\{x_{pj} \le c_j\}$ is γ_{il} but for the region $\{x_{pj} > c_j\}$ one has to distinguish between $\{x_{pj} > c_j, x_{ps} \le c_s\}$ with item difficulty $\gamma_{il}^{(n)}$ and $\{x_{pj} > c_j, x_{ps} > c_s\}$ with item difficulty $\gamma_{ir}^{(n)}$. The corresponding test refers to the null hypothesis H_0: $\gamma_{il}^{(n)} - \gamma_{ir}^{(n)} = 0$.

The trees that are obtained are trees for specific items, namely the items that were selected to carry DIF. If an item is never selected it is considered as compatible with the Rasch model. It should be mentioned that in the description of the splits it was assumed that variables are metrically scaled or ordinal. If variables are categorical and nominally scaled the splitting rule has to be modified. For categorical variables without ordering $x_j \in \{1, \ldots, k_j\}$, the partition has the form $A \cap S, \quad A \cap \bar{S}$, where S is a non-empty subset $S \subset \{1, \ldots, k_j\}$ and $\bar{S} = \{1, \ldots, k_j\} \setminus S$ is the complement.

DIF detection by item-focused trees can also be used in other binary models and polytomous models. An extension to the partial credit model has been considered by Bollmann et al. (2018).

Example 10.4 Knowledge Data (Tutz and Berger, 2016a)
Let us consider the same data as in Example 10.3. When using item focused recursive partitioning 21 of the 45 items show DIF. As noted before this not surprising because the questions of the quiz were not chosen very carefully to avoid DIF. Altogether the algorithm performs 33 splits until further splits are not significant at significance level $\alpha = 0.05$ (for details of the test see Tutz and Berger (2016a)). The first ten splits all refer to the covariate gender, so the strongest effects were found for the difference between males and females. No significant splits were found for the variable elite. The difficulties of the items seem not to depend on the elite status of the university, which is in line with the weak elite effects seen in Example 10.1. The three items with the strongest effects, which were found in the first iterations of the algorithm, were the following:

- 19: Who is this?—Picture of Dieter Zetsche, CEO of Mercedes-Benz
- 43: Which kind of bird is this?—Blackbird
- 40: What is also termed Trisomy 21?—Down syndrome.

The resulting trees for these items 19, 43, and 40 are shown in Fig. 10.4. For each item one can see how the difficulty of the item depends on the characteristic of certain variables. The estimated item difficulties are given in each leaf of the trees, which represent the identified subgroups. For example, in item 19 (recognition of Dieter Zetsche as CEO of Mercedes Benz) the difficulty for females (gender $= 0$) is 2.665 while for males (gender $= 1$) it is distinguished between students who frequently read Spiegel online (spon $= 7$) with an item difficulty of 0.126 and a much larger item difficulty of 1.155 for students who read it less regularly (spon ≤ 6). The other two items show DIF only for gender. Both items concerning the recognition of birds and knowledge of genetic diseases are easier to solve for females. It is also seen that item 19 is much harder to solve than the other two items.

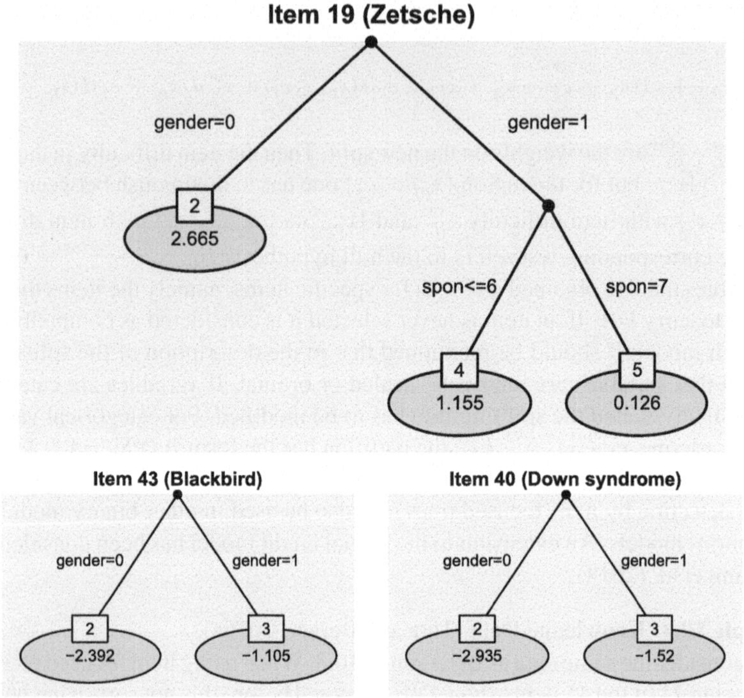

Fig. 10.4 Trees for Items 19, 43, and 40 of the general knowledge test. The parameters of item-difficulty are given for each subgroup represented by the leaves of the trees

Fitting of item-focussed trees can be obtained by using the package *DIFtree* (Berger, 2020), see also (Tutz and Berger, 2016a; Berger and Tutz, 2016):

```
library(DIFtree)
library(psychotree)
data(SPISA)
SP <- as.data.frame(SPISA[,1])  #responses
C<- SPISA[,-1]  #covariates
##### uniform DIF, Rasch model
mf<-DIFtree(SP , C , model = c("Rasch"), type = c("udif"),
  alpha = 0.05, nperm = 1000, trace = FALSE,penalize = FALSE)
print(mf)
plot(ml1,item=19)
```

Item-Focused Trees for Uniform DIF

The item-focused trees considered in the previous section work for parametric models as the Rasch model or extensions thereof. An alternative way to investigate DIF can be derived by using the logistic regression framework considered in (10.1) but with tree components instead of the linear models used in classical logistic regression. It allows to investigate uniform as well as no-uniform DIF.

The logistic model for uniform DIF detection has the form $\log(P(Y_{pi} = 1|S_p, \boldsymbol{x}_p)/P(Y_{pi} = 0|S_p, \boldsymbol{x}_p)) = \eta_{pi}$, where $\eta_{pi} = \beta_{0i} + S_p\beta_i + \boldsymbol{x}_p^\top\gamma_i$ is linear in $\boldsymbol{x}_p$. An item-focused version is obtained if the linear term $\boldsymbol{x}_p^\top\gamma_i$ is replaced by a tree (Berger and Tutz, 2016).

When growing a tree the first split with regard to the jth variable and corresponding split point c_j is obtained by fitting the model with predictor

$$\eta_{pi} = S_p\beta_i + [\gamma_{il}^{[1]} I(x_{pj} \le c_j) + \gamma_{ir}^{[1]} I(x_{pj} > c_j)], \tag{10.7}$$

where $I(\cdot)$ denotes the indicator function with $I(a) = 1$ if a is true and $I(a) = 0$ otherwise. The parameter $\gamma_{il}^{[1]}$ denotes the intercept in the left node ($x_{pj} \le c_j$) and $\gamma_{ir}^{[1]}$ the intercept in the right node ($x_{pj} > c_j$). For example, one split with regard to the binary covariate gender yields the intercepts

$$\gamma_{il}^{[1]} = \gamma_{i,male} \text{ for males and } \gamma_{ir}^{[1]} = \gamma_{i,female} \text{ for females.}$$

The main difference to the use of a linear predictor is that the two subgroups of interest are not predefined but determined by a split in variable j at split-point c_j. To determine the first split one examines all the null hypotheses $H_0 : \gamma_{il}^{[1]} = \gamma_{ir}^{[1]}$. If H_0 cannot be rejected for any combination of variable and split point the item is considered to be free of DIF. In the algorithm likelihood ratio tests are used to examine the null hypotheses. In the very first step one chooses the combination of item, variable, and split point with the smallest p-value of the corresponding test. If a significant effect is found the first split into left and right node is carried out for the selected item.

One further split, for example in the right node ($x_{pj} > c_j$), with regard to the sth variable at split point c_s yields the two daughter nodes $I(x_{pj} > c_j)I(x_{ps} \le c_s)$ and $I(x_{pj} > c_j)I(x_{ps} > c_s)$. The new nodes are both defined by the product of two indicator functions. In general each node can be represented by a product of several indicator functions, namely

$$node(\boldsymbol{x}_p) = \prod_{b=1}^{B} I(x_{pj_b} > c_{j_b})^{a_b} I(x_{pj_b} \le c_{j_b})^{1-a_b},$$

where B is the total number of indicator functions or branches, c_{j_b} is the selected split point in variable j_b and $a_b \in \{0, 1\}$ indicates which of the indicator functions, below or above the threshold, is involved. The resulting predictor of the model for item i after several splits with terminal nodes $\ell = 1, \ldots, L_i$ is then given by

$$\eta_{pi} = S_p\beta_i + \sum_{\ell=1}^{L_i} \gamma_{i\ell}\, node_{i\ell}(\boldsymbol{x}_p) = S_p\beta_i + tr_i(\boldsymbol{x}_p). \tag{10.8}$$

where $tr_i(\boldsymbol{x}_p)$ is the tree component containing subgroup-specific intercepts represented by the terminal nodes $node_{i\ell}(\boldsymbol{x}_p)$. The proposed algorithm yields an individual tree for each item that was selected to carry DIF. If an item is never chosen for splitting it is assumed to be free of DIF, and the fitted "tree" is a constant $tr_i(\boldsymbol{x}_p) = \beta_{0i}$. For details see Berger and Tutz (2016).

Item-Focused Trees for Non-uniform DIF

The logistic model for non-uniform DIF detection has the form $\log(P(Y_{pi} = 1|S_p, \boldsymbol{x}_p)/\ P(Y_{pi} = 0|S_p, \boldsymbol{x}_p)) = \eta_{pi}$, where $\eta_{pi} = \beta_{0i} + S_p\beta_i + M_p\gamma_i + S_p M_p\alpha_i$, in which an interaction term is included. An item-focused version is obtained if the linear term $\boldsymbol{x}_p^\top\gamma_i$ is replaced by a tree (Berger and Tutz, 2016).

As in the extended linear logistic model several strategies can be used to detect DIF. The first strategy is to test for both types of DIF effects simultaneously. For item-focused trees that means that at least one split is performed in one of the components. When selecting the optimal split in each step of the algorithm, one has to consider all combinations of item, variable, split point, and component with regard to intercept and slope. The final model consists of one or two separate trees, one referring to the intercept and one referring to the slope. The resulting tree is given by

$$\eta_{pi} = tr_i(\boldsymbol{x}_p) + tr_i(S_p, \boldsymbol{x}_p), \tag{10.9}$$

where $tr_i(\boldsymbol{x}_p)$ is the tree component containing subgroup-specific intercepts and $tr_i(S_p, \boldsymbol{x}_p)$ is the tree component containing subgroup-specific slopes. In contrast to the tree in model (10.8) for uniform DIF now one has two possible trees. If there is only a significant effect in one of the two components a constant $tr_i(\boldsymbol{x}_p) = \beta_{0i}$ or $tr_i(S_p, \boldsymbol{x}_p) = S_p\beta_i$ is fitted in the other component.

In comparison to the classical and extended *logistic* method, the tree based model has two advantages: The obtained tree(s) distinguish between items with uniform and non-uniform DIF. The trees themselves show which form of DIF is present. Thus both types of DIF can be detected simultaneously within one fitting procedure. The obtained tree(s) identify the variables that induce uniform and/or non-uniform DIF. In particular, both types of DIF can be caused by different variables.

The second strategy is to explicit test for non-uniform DIF. For item-focused trees it means that a significant split in the *slope* component is found. Consequently, during estimation only the models with *simultaneous splits* in the intercepts and the slopes are considered as potential candidates. Therefore, one split in item i with regard to variable j corresponds to the model with predictor

$$\eta_{pi} = [\gamma_{il}^{[1]} I(x_{pj} \le c_j) + \gamma_{ir}^{[1]} I(x_{pj} > c_j)] + S_p\ [\alpha_{il}^{[1]} I(x_{pj} \le c_j) + \alpha_{ir}^{[1]} I(x_{pj} > c_j)], \tag{10.10}$$

which contains two intercepts $(\gamma_{il}^{[1]}, \gamma_{ir}^{[1]})$ and two slopes $(\alpha_{il}^{[1]}, \alpha_{ir}^{[1]})$ with respect to the same subgroups. To select the optimal split and to determine the splitting decision one compares the likelihoods of model (10.7) and (10.10). The procedure is continued in each step of the algorithm, considering all combinations of item, variable, and split point. If non-uniform DIF is present, the final model consists of two trees containing subgroup-specific intercepts and subgroup specific slopes that are determined by the same splits.

Example 10.5 CTB Science data (Berger and Tutz, 2016)
We consider a data set from CTB-McGraw Hill. For a description of the original data, see also De Boeck and Wilson (2004). The data includes the results of 1500 grade 8 students from 35 schools. The students had to respond to 76 items, measuring different objectives and subskills related to mathematics and science. In our investigation we restrict to the 25 multiple-choice items from subject area science. To test for DIF in these items we consider the three covariates gender (male: 0, female: 1), type of the school (1: catholic, 2: private, 3: public) and size of the school (number of students in hundreds). The methods are abbreviated in the following way:

- *UDIF* means testing for uniform DIF in the extended logistic model, that is, $H_0 : \gamma_i = 0$, given model (10.2).
- *UDIF-trees* refers to item-focused trees for uniform DIF.
- *DIF* means simultaneous tests for uniform and non-uniform DIF, $H_0 : \gamma_i = \alpha_i = 0$, given model (10.4) for logistic regression.
- *DIF-trees* refers to item-focused trees for non-uniform DIF (simultaneous tests),
- *NUDIF* means tests for non-uniform DIF, $H_0 : \alpha_i = \mathbf{0}$, given model (10.4) for logistic regression.
- *NUDIF-trees* refers to the corresponding item-focussed tree.

When using trees to detect uniform DIF 14 of 25 items were identified as DIF items. Altogether the algorithm performs 27 splits until further splits are no longer significant. With three covariates, each permutation test is performed at local significance level $0.05/3$. The p-value in the 28th iteration was 0.02 and thus not significant on level $0.01\overline{6}$. All splits refer to covariates type and size, whereas no significant splits were found for variable gender. There does not seem to be any difference between males and females.

The trees for three selected items are given in Fig. 10.5. In item 10 DIF is induced by size and one has to distinguish between three subgroups. The item is easiest for students in small schools (size ≤ 400) but most difficult for students in medium-sized schools ($400 <$ size ≤ 900). Item 21 is easier for students in a catholic or private school (type ≤ 2) compared to students in public schools (type $= 3$). An interesting partition is received for item 25. For all students in a catholic school (type $= 1$) the question is very difficult. By contrast the question is easier for all students in a private or public school (type > 1), in particular for those in medium-sized schools ($500 <$ size ≤ 1000).

An overview of the detected DIF items by the six evaluated models is given in Table 10.2. It shows only items that were found to be DIF items by at least one of the models. Within the *DIF* framework (second column) eleven DIF items are identified. These are the same items as with the restricted model for uniform DIF discussed

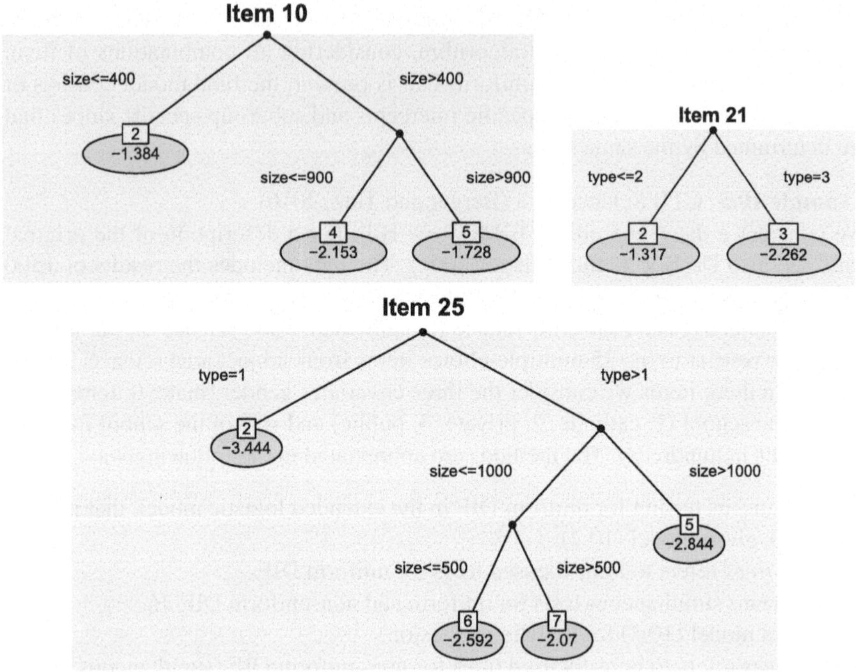

Fig. 10.5 Trees of items 10, 21, and 25 of the CTB data using the model for uniform DIF. Estimated intercepts γ_{il} are given in each leaf of the trees. Reprinted with permission from Berger and Tutz (2016)

above, but without item 6, 15, and 17. There are three items that are classified as non-uniform DIF items by the more general model. Here, for example in item 21 the split regarding the type of school is not performed in the intercept but in the slope component. According to the model testing for non-uniform DIF (third column) the two items 13 and 21 carry non-uniform DIF. In contrast to item 13, item 21 is also detected within the *UDIF* and *DIF* framework.

The comparison to the linear logistic approach shows a strong overlap. Within the *UDIF* framework (first and fourth column) there is a agreement in nine items. In the *DIF* framework this is the case for eight items. However it should be mentioned, that the linear logistic approach does not distinguish between uniform and non-uniform DIF. When testing for non-uniform DIF (sixth column) one obtains four significant results. In contrast to items 1 and 11, items 13 and 21 are also found to carry DIF. Item 21 is the only item that shows DIF according to all six models and four items are only identified as DIF items by one of the six models. For more details see Berger and Tutz (2016).

For an overview on available packages to investigated DIF see Chap. 12.

Table 10.2 Comparison of detected DIF items of the CTB data using IFT and the extended Logistic approach for uniform and non-uniform DIF

Item	Item focussed trees				Extended logistic		
	UDIF-trees	DIF-trees		NUDIF-trees	UDIF	DIF	NUDIF
21	×	×	(non)	×	×	×	×
3	×	×	(u)		×	×	
4	×	×	(u)		×	×	
8	×	×	(u)		×	×	
9	×	×	(u)		×	×	
14	×	×	(non)		×	×	
16	×	×	(non)		×	×	
25	×	×	(u)		×	×	
11					×	×	×
13				×		×	×
19	×	×	(u)		×		
5	×	×	(u)				
10	×	×	(u)				
24					×	×	
1							×
6	×						
15	×						
17	×						

Source Reprinted with permission from Berger and Tutz (2016)

10.4 Fitting of Models

The following packages can be used to investigate differential item functioning:

- R package *difR*: Investigates DIF linked to group membership, includes many methods, among them, generalized Lord, generalized Mantel-Haenszel, logistic regression, Lord's chi-square test, Mantel-Haenszel, Raju's area, see Magis et al. (2010).
- R package *DIFlasso*: Penalized Ml estimation for Rasch models, allows the simultaneous selection of items and variables that induce differential item functioning (Tutz and Schauberger, 2015).
- R package *GPCMlasso*: Allows to identify DIF in Generalized Partial Credit Models. Several variables can be considered simultaneously. For each pair between variable and item, a parametric DIF effect is introduced which indicates DIF if the respective parameter is selected (Schauberger and Mair, 2020).

- R package *DIFtree*: Item focussed recursive partitioning for the simultaneous selection of items and variables that induce differential item functioning (DIF) in dichotomous or polytomous items (Tutz and Berger, 2016a; Berger and Tutz, 2016).
- R package *psychotree*: Rasch trees as proposed by Strobl et al. (2015).

Chapter 11
Explanatory Item Response Models

In the explanatory item response modeling (EIRM) approach (De Boeck and Wilson, 2004, 2016) item responses are seen as dependent variables to be explained by independent variables, which can be subject-specific but also item-specific variables. The focus of the modeling is on the impact of the explanatory variables and person parameters are considered as accounting for the heterogeneity of responses. The approach allows to investigate which variables determine the responses and is related to differential item functioning considered in Chap. 10.

Basic IRT models assume that latent trait are unidimensional and not affected by covariates. If one suspects that covariates may modify the response behaviour it can be tested by including them explicitly in the explanatory term. We consider in particular ordinal models, binary models are covered implicitly as special cases.

11.1 Subject-Specific Covariates in Ordinal Models

In the following let $\boldsymbol{x}_p^T = (x_{p1}, \ldots, x_{pm})$ denote a subject-specific vector of covariates that is assumed to potentially modify the responses. The responses are polytomously scored such that $Y_{pi} \in \{0, 1, \ldots, k\}$, $p = 1, \ldots, P$, $i = 1, \ldots, I$, with $\{0, 1, \ldots, k\}$ representing the ordering of categories.

11.1.1 The Partial Credit Model with Covariates

The basic partial credit model (PCM) has the form

$$\log\left(\frac{P(Y_{pl} = r)}{P(Y_{pi} = r - 1)}\right) = \theta_p - \delta_{ir}, \quad r = 1, \ldots, k, \tag{11.1}$$

G. Tutz, *A Short Guide to Item Response Theory Models*, Statistics for Social and Behavioral Sciences, https://doi.org/10.1007/978-3-031-87271-6_11

with person parameters θ_p and item parameters δ_{ir}. In a PCM with covariates the person parameter θ_p is replaced by $\theta_p + \boldsymbol{x}_p^T\gamma$ yielding

$$\log\left(\frac{P(Y_{pi}=r|\boldsymbol{x}_p)}{P(Y_{pi}=r-1|\boldsymbol{x}_p)}\right) = \theta_p + \boldsymbol{x}_p^T\gamma - \delta_{ir}, \quad r = 1,\ldots,k. \tag{11.2}$$

The parameter $\boldsymbol{\gamma}^T = (\gamma_1, \ldots, \gamma_m)$ contains the effects of covariates on the response behavior. The strength of the effect is seen by considering the *local odds*

$$\kappa_{pir}(\boldsymbol{x}_p) = P(Y_{pi}=r|\boldsymbol{x}_p)/P(Y_{pi}=r-1|\boldsymbol{x}_p).$$

If the jth component of the vector $\boldsymbol{x}_p$ increases by one unit while all other variables are kept fixed the local odds change by the factor e^{γ_j}. More formally, one has

$$\frac{\kappa_{pir}(x_{p1},\ldots,x_{pj}+1,\ldots,x_{pm})}{\kappa_{pir}(x_{p1},\ldots,x_{pm})} = e^{\gamma_j}. \tag{11.3}$$

For $\gamma_j > 0$ all local odds increase if the jth variable increases and responders with larger covariate values tend to choose higher categories. If $\gamma_j < 0$ lower categories are to be expected. A more detailed description of the model is given in Schauberger and Tutz (2022).

Example 11.1 Fears data
As an illustrating example, we consider again the fears data. The participants were asked: "How afraid are you due to the ..."(1) refugee crisis? (2) global climate change? (3) international terrorism? (4) globalization? (5) use of nuclear energy? with answers measured on Likert scales from 1 (not afraid at all) to 7 (very afraid). A partial credit model is fitted including the (standardized) covariates gender (1: female; 0: male), age, EastWest (1: Eastern German countries/former GDR, 0: Western German countries/former FRG) and Abitur (High school degree for the admission to the university, 1: yes, 0: no).

The model can be fitted by using the package *MultOrdRS*, which yields marginal log-likelihood estimates under the assumption that person parameters are normally distributed.

```
library("MultOrdRS")
data(GLES17)
## scale data
GLES17[,7:11] <- scale(GLES17[,7:11])
## define formula
f.GLES <-
  as.formula(cbind(RefugeeCrisis, ClimateChange, Terrorism,
  Globalization,  NuclearEnergy) ~
  Age + Gender + Unemployment + EastWest + Abitur)
## fit adjacent categories model
adj0 <-
 multordRS(f.GLES, data = GLES17, control = ctrl.multordRS(RS = FALSE))
```

The resulting effects of covariates are:

```
Location effects:
              Estimate    Std.Err Z value Pr(>z)
Age          0.1182001  0.0118885  9.9424 <2e-16 ***
Gender       0.1145677  0.0116061  9.8714 <2e-16 ***
Unemployment 0.0002550  0.0116006  0.0220 0.9825
EastWest    -0.0083031  0.0115305 -0.7201 0.4715
Abitur      -0.1093216  0.0118048 -9.2608 <2e-16 ***

Estimates of random effects (co-)variance Sigma
Intercept
0.1660647

Log-Likelihood: -17751.64 with 36 degrees of freedom
```

It is seen that age, gender, and Abitur have highly significant effects on the responses. Older people and females tend to have higher levels of fear, higher education tends to reduce fear. □

In a next step one can include response style effects, which when neglected might yield biased estimates of location effects. More concrete, the tendency to middle or extreme categories is included. The corresponding model is

$$\log\left(\frac{P(Y_{pi}=r|\boldsymbol{x}_p)}{P(Y_{pi}=r-1|\boldsymbol{x}_p)}\right)=\theta_p+\boldsymbol{x}_p^T\boldsymbol{\gamma}+(k/2-r+0.5)(\alpha_{op}+\boldsymbol{x}_p^T\boldsymbol{\alpha})-\delta_{ir},\quad r=1,\ldots,k, \tag{11.4}$$

which is an extension of the PCM considered in Chap. 5. The parameters in γ are referred to as location effects since they represent the tendency to higher or lower response categories. The parameters $\boldsymbol{\alpha}$ represent the response style effects. The effect of increasing one component of the explanatory variable by one unit is seen when considering the effect on the local odds,

$$\frac{\kappa_{pir}(x_{p1},\ldots,x_{pj}+1,\ldots,x_{pm})}{\kappa_{pir}(x_{p1},\ldots,x_{pm})}=e^{\gamma_j}e^{(k/2-r+0.5)\alpha_j}. \tag{11.5}$$

If α_j is positive an increase in the j-the variable means that the tendency to middle categories increases. For negative values the tendency to extreme categories increases. For illustration, we consider $k=4$. Then, the factors $e^{(k/2-r+0.5)\alpha_j}$, which distinguish between category r and $r-1$, are given by

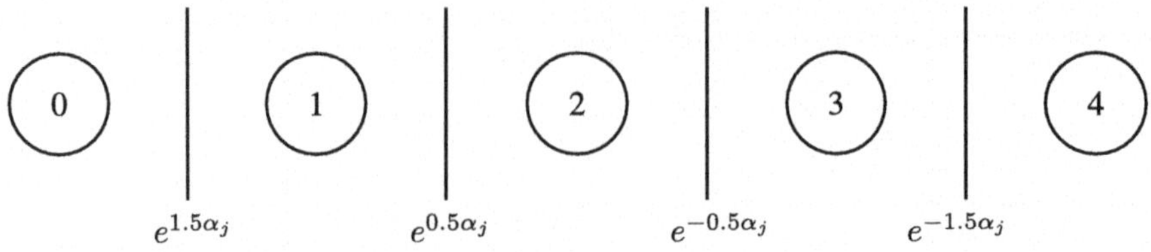

It is seen that positive values of α_j increase the probabilities in middle categories. In (11.5) location effects and response style effects are separated. A variable can also have just one effect, for example $\gamma_j > 0$ and $\alpha_j = 0$ if there is a positive location effect but no response style effect.

Marginal estimates of effects are obtained by assuming a common normal distribution of (θ_p, α_{op}) with mean zero and covariance matrix $\boldsymbol{\Sigma}$.

Example 11.2 Fears data
Let us consider again the fears data considered in Example 11.1. Fitting of the model is obtained by:

```
adj <-
 multordRS(f.GLES, data = GLES17, control = ctrl.multordRS(cores = 6))
plot(adj, alpha = 0.05, CIfactor = 0.9,cex.axis=1.5,cex.lab=1.4,
 cex.main=1.5)
```

Location and response styles effects are

```
Location effects:
              Estimate   Std.Err Z value Pr(>z)
Age           0.111339  0.012104  9.1987 <2e-16 ***
Gender        0.114617  0.011842  9.6791 <2e-16 ***
Unemployment -0.001552  0.011652 -0.1332 0.8940
EastWest     -0.010284  0.011701 -0.8789 0.3795
Abitur       -0.106457  0.011974 -8.8909 <2e-16 ***

Response style effects:
               Estimate    Std.Err Z value    Pr(>z)
Age          -0.0522740  0.0097198 -5.3781 7.529e-08 ***
Gender       -0.0053790  0.0094428 -0.5696    0.5689
Unemployment -0.0123519  0.0092311 -1.3381    0.1809
EastWest     -0.0115590  0.0093146 -1.2410    0.2146
Abitur        0.0141111  0.0096106  1.4683    0.1420

Estimates of random effects (co-)variance Sigma
            Intercept   RespStyle
Intercept 0.156045955 0.002263342
RespStyle 0.002263342 0.070299133

Log-Likelihood: -17597.26 with 43 degrees of freedom
```

It is seen that the location effects remain very stable when response style effects are included. The location parameter estimates are almost the same as in Example 11.1. However, this does not hold in general. It is seen in the present example since the dispersion effects are not very strong, only one variable shows a distinctly significant effect. Older respondents have a significant tendency to more extreme categories than younger persons.

The estimates of the parameters of the random effects show that the variance of the location parameters is larger than that of the response style parameters. The correlation between the two effects is very small (correlation coefficient 0.02). □

A visualization of the parameter estimates can be obtained by using star plots (Tutz and Berger, 2016b). Figure 11.1 shows the tuples $(e^{\hat{\gamma}}, e^{\hat{\alpha}})$. The first value, $e^{\hat{\gamma}}$, represents the location effect on the odds. For values larger than one high response categories are favored, for values smaller than one low response categories are favored. The second value, $e^{\hat{\alpha}}$, represents the response style effect, for values larger than one there is a tendency to middle categories, for values smaller than one there is a stronger tendency to extreme categories. Also pointwise 95% confidence intervals are included, represented by stars with the horizontal and vertical lengths corresponding to the confidence intervals of $e^{\hat{\gamma}}$ and $e^{\hat{\alpha}}$, respectively. If stars cross the horizontal or/and vertical lines the corresponding effects can not be considered as significant. For example, the star for unemployment crosses both lines since both effects are not significant, while the star for Age is far away from both lines, indicating strong significance.

An alternative approach to include variables propagated by Zhou (2019) starts with the partial credit model (11.1) and assumes that the distribution of the latent trait is a normal distribution $N(\mu_p, \sigma_p^2)$ with the mean and variance determined by

$$\mu_p = \boldsymbol{x}_p^T \gamma \quad \log(\sigma_p^2) = \boldsymbol{x}_p^T \boldsymbol{\beta}.$$

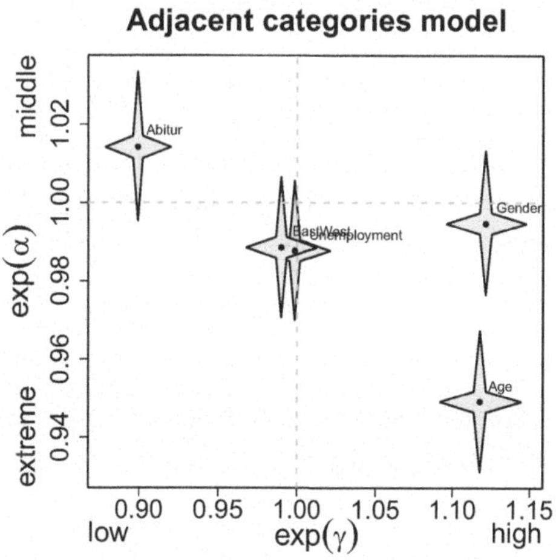

Fig. 11.1 Star plot for fears data, partial credit model with response style

The first term specifies how the mean of the latent trait depends on the covariates, the second term specifies that the variance of the latent trait is a function of covariates. It is a hierarchical latent trait model, which on the first level assumes a latent trait model and on the second level specifies the dependence on covariates.

If in the model the variance is constant, that is $\log(\sigma_p^2) = c$, the model is equivalent to the model with covariates but without response style (11.2). In the heterogeneous model with varying variance, the model is not equivalent to the model with response style (11.4). Although the former model explicitly models dispersion of the latent trait in the term $\log(\sigma_p^2) = \boldsymbol{x}_p^T \boldsymbol{\beta}$ the effect of covariates is very similar to that of the model with response style (11.4), covariates determine if there is a tendency to middle or extreme categories. There is one distinct difference between the models concerning the modelling of the tendency to middle or extreme categories. Zhou's model uses just one latent trait, it is a one-dimensional person parameter model and does not include an individual response style parameter α_{0p}. It models the dependence of variance on covariates without accounting for the heterogeneity of individuals, in contrast to model (11.4), which uses two latent dimensions, one for the location and one for the response style. More applications and comparisons of modeling approaches are found in Mauerer and Tutz (2024) and Schauberger and Tutz (2022).

11.1.2 The Graded Response Model with Covariates

The basic graded response model (GRM) has the form

$$P(Y_{pi} \geq r) = F(\theta_p - \delta_{ir}), \quad r = 1, \ldots, k,$$

with person parameters θ_p and item parameters δ_{ir}. In a GRM with covariates the person parameter θ_p is replaced by $\theta_p + \boldsymbol{x}_p^T \gamma$ yielding

$$P(Y_{pi} \geq r) = F(\theta_p + \boldsymbol{x}_p^T \gamma - \delta_{ir}), \quad r = 1, \ldots, k, \tag{11.6}$$

If $F(.)$ is the logistic function the interpretation of parameters is very similar to the interpretation in the partial credit model but instead of local odds one has to use cumulative odds $\kappa_{pir}(\boldsymbol{x}_p) = P(Y_{pi} \geq r|\boldsymbol{x}_p)/P(Y_{pi} < r|\boldsymbol{x}_p)$. For these odds one has

$$\frac{\kappa_{pir}(x_{p1}, \ldots, x_{pj} + 1, \ldots, x_{pm})}{\kappa_{pir}(x_{p1}, \ldots, x_{pm})} = e^{\gamma_j}.$$

For the graded response model the inclusion of response styles is harder to obtain than for the PCM. The predictor $\eta_{pr} = \theta_p + \boldsymbol{x}_p^T \gamma + (k/2 - r + 0.5)(\gamma_p +$

$\boldsymbol{x}_p^T\boldsymbol{\alpha}) - \delta_{ir}$, which was used in the extended PCM, can not be used for normally distributed parameters γ_p since the restriction $P(Y_{pi} \geq r) \geq P(Y_{pi} \geq r+1)$ tends to be violated. However, a slightly modified version with a similar term can be used after reparameterization, for details, see Schauberger and Tutz (2022).

Also for the cumulative model one can use the hierarchical model propagated by Zhou (2019). Starting from the basic cumulative model the assumptions $\mu_p = \boldsymbol{x}_p^T\gamma$, $\log(\sigma_p^2) = \boldsymbol{x}_p^T\boldsymbol{\beta}$ can be used without any reparameterization. In the following example also Zhou's model is considered.

Example 11.3 Fears data
For the fears data considered in Example 11.1, the graded response model ($\alpha_i = 1$) can be fitted and plotted with the MultOrdRS package by using

```
cum <- multordRS(f.GLES, data = GLES17, control =
             ctrl.multordRS(cores = 6),model = "cumulative")
plot(cum, alpha = 0.05, CIfactor = 0.85,cex.axis=1.5,cex.lab=1.8,
             cex.main=1.5,main ="Cumulative model")
```

yielding

```
Location effects:
                Estimate   Std.Err Z value Pr(>z)
Age            -0.305864  0.032915 -9.2924 <2e-16 ***
Gender         -0.316780  0.032238 -9.8264 <2e-16 ***
Unemployment  0.011108  0.032255  0.3444 0.7306
EastWest       0.034802  0.032221  1.0801 0.2801
Abitur         0.292678  0.032717  8.9456 <2e-16 ***

Response style effects:
                Estimate    Std.Err Z value    Pr(>z)
Age           -0.0505305  0.0129955 -3.8883 0.0001009 ***
Gender         0.0125918  0.0126354  0.9966 0.3189802
Unemployment -0.0166719  0.0129461 -1.2878 0.1978193
EastWest      -0.0259639  0.0126319 -2.0554 0.0398385 *
Abitur         0.0075586  0.0128710  0.5873 0.5570280

Estimates of random effects (co-)variance Sigma
          Intercept  RespStyle
Intercept  1.207621 -0.1633100
RespStyle -0.163310  0.1607543

Log-Likelihood: -17508.18 with 43 degrees of freedom
```

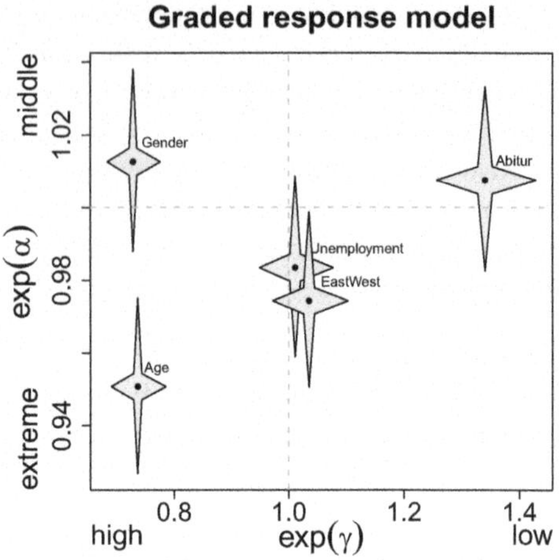

Fig. 11.2 Star plot for fears data, graded response model with response style

It is seen that the same effects are significant as in the adjacent categories model. Figure 11.2 shows the corresponding star plot with 0.95 confidence intervals. In the picture the location effects are reversed as compared to Fig. 11.1 since the program uses the alternative form of the graded response model $P(Y_{pi} \leq r) = F(\eta_{pir})$. The graded response model shows better fit than the corresponding partial credit model.

As an alternative Zhou's cumulative model is considered, which can be fitted by using:

```
library("MultOrdRS")
data(GLES17)
## scale data
GLES17[,7:11] <- scale(GLES17[,7:11])

yn<-GLES17[,c(1,2,3,4,6)]
xn<-model.matrix( ~ Age + Gender + Unemployment + EastWest + Abitur,
      GLES17)
zn<-model.matrix( ~ Age + Gender + Unemployment + EastWest + Abitur,
      GLES17)

fitcum<-hgrm(yn, xn)
fitcum

fitcumdisp<-hgrm(yn, xn, zn)
fitcumdisp
```

The first model is the cumulative model that uses only $\mu_p = \boldsymbol{x}_p^T\gamma$ and yields the following output:

```
Mean Regression:
                    Estimate Std_Error z_value p_value
x(Intercept)           0.000     0.094   0.000   1.000
xAge                   0.294     0.033   9.023   0.000
xGender                0.275     0.032   8.489   0.000
xUnemployment         -0.009     0.024  -0.356   0.722
xEastWest              0.031     0.028   1.130   0.258
xAbitur               -0.299     0.035  -8.418   0.000

Variance Regression:
          Estimate Std_Error z_value p_value
zx1              0     0.133       0       1

Log Likelihood: -17613.76
```

Comparison with the fitting of the corresponding partial credit model (Example 11.1) shows that the same variables have significant effects, however, the cumulative model fits better. The model that includes the additional effect $\log(\sigma_p^2) = \boldsymbol{x}_p^T\boldsymbol{\beta}$, which can be seen as a response style effect, yields the output

```
Mean Regression:
                    Estimate Std_Error z_value p_value
x(Intercept)           0.000     0.095   0.000   1.000
xAge                   0.309     0.034   9.101   0.000
xGender                0.282     0.033   8.567   0.000
xUnemployment         -0.017     0.032  -0.534   0.593
xEastWest              0.031     0.028   1.114   0.265
xAbitur               -0.303     0.035  -8.691   0.000

Variance Regression:
                    Estimate Std_Error z_value p_value
z(Intercept)           0.000     0.135   0.000   1.000
zAge                   0.154     0.045   3.421   0.001
zGender                0.027     0.046   0.589   0.556
zUnemployment          0.071     0.053   1.338   0.181
zEastWest              0.009     0.046   0.195   0.845
zAbitur               -0.125     0.048  -2.602   0.009

Log Likelihood: -17602.23
```

It is seen that in the mean regression term the same three variables as in the corresponding model fitted by MultOrdRS are significant. In the variance term age and Abitur are significant whereas in the model fitted by MultOrdRS age and East-West are significant. The model with response style shows better fit (log-likelihood

−17508.18) than Zhou's model (log-likelihood −17602.23), however, the former includes an individual parameter for the response style, which Zhou's model does not. □

11.2 Covariates in the Threshold Model

The thresholds model allows for different kinds of responses including ordinal or count data. In the following the inclusion of covariates for this general class of models is considered. Also a more general form of covariate effects that allows for an interaction between covariates and items is considered.

Let again $\boldsymbol{x}_p$ be a person-specific vector of covariates. In a threshold model with covariates the person parameter θ_p is replaced by $\theta_p + \boldsymbol{x}_p^T \gamma_i$ yielding

$$P(Y_{pi} > y|\theta_p, \alpha_i, \delta_i(.), \boldsymbol{x}_p) = F(\alpha_i(\theta_p + \boldsymbol{x}_p^T \gamma_i - \delta_i(y))), \tag{11.7}$$

where the parameter $\gamma_i^T = (\gamma_{i1}, \ldots, \gamma_{im})$ is item-specific and represents the effect on the response in item i. It means that the effect of covariates can vary across items, for some items there might be no effect at all while the performance on other items is strongly modified by covariates.

Model (11.7) is a rather general multivariate regression model with heterogeneity. It can be seen as a generalized random effects model, but with much weaker assumptions on the distribution of the response variables than in the classical linear mixed model considered, for example, by Goldstein (1987); Searle et al. (1992). More general than in the previously considered ordinal models the effects of explanatory variables may vary across items. An extensive treatment of the mixed thresholds model is found in Tutz (2025).

The model has many parameters since each covariate contributes as many parameters as items are used. A natural question is if the effects can be simplified. Instead of varying over items one might assume that effects are *global*, that is, not varying over items. This has been implicitly assumed in the modeling of covariate effects in the partial credit and graded response model. It can be tested if effects are global for the jth variable by using a likelihood ratio test for the hypotheses

$$H_0 : \gamma_{1j} = \cdots = \gamma_{Ij} \text{ against } H_1 : \gamma_{sj} \neq \gamma_{lj} \text{ for at least one pair } (s, l).$$

Example 11.4 Fears data

As an illustrating example, we consider again the fears data considered in Example 4.5. We fitted a discrete thresholds with logistic response function and logit difficulty function including covariates, gender (1: female; 0: male), standardized age in decades, EastWest (1: Eastern German countries/former GDR, 0: Western German countries/former FRG) and Abitur (High school degree for the admission to the university, 1: yes, 0: no). Table 11.1 shows the estimates of item parameters for the basic

Table 11.1 Parameter estimates for the fears data with logit difficulty function, logistic response function without covariates and with covariates, z-values of parameter estimates of covariate parameters are given in the lower part, variable age was standardized

		Model without covariates					
	Item	Intercepts	Slopes	Age	Gender	EastWest	Abitur
1	Refugee	−0.255	1.575				
2	Climate change	−1.391	1.834				
3	Terrorism	−2.153	1.644				
4	Globalization	0.225	1.789				
5	Nuclear energy	−0.837	1.655				
	Log-lik	−1728.913					
		Model with item-specific effects					
1	Refugee	−0.341	1.653	0.151	0.606	0.392	−1.388
2	Climate change	−1.140	1.858	0.013	1.061	−0.594	−0.062
3	Terrorism	−2.062	1.795	0.393	1.329	0.321	−1.207
4	Globalization	0.716	1.862	0.195	1.020	0.725	−0.719
5	Nuclear energy	−0.922	1.641	0.254	0.379	−0.416	−0.300
	Log-lik	−1684.871					
		z-values					
1	Refugee			0.937	1.883	1.179	−4.002
2	Climate change			0.079	3.313	−1.798	−0.186
3	Terrorism			2.354	3.935	0.939	−3.468
4	Globalization			1.236	3.184	2.206	−2.108
5	Nuclear energy			1.565	1.189	−1.253	−0.892
		Model with global effects					
	Log-lik	−1715.82		0.230	1.030	0.290	−0.545

model without covariates and the model with covariates (discrete response, $\alpha_i = 1$, logistic response function, logit difficulty function). The parameters given are the intercepts, the slopes of the difficulty function $\delta(y) = \delta_{0i} + \delta_i \log((y - a)/(b - y))$ and the item-specific covariate effects. It is seen that all items show significant covariate effects for at least one of the covariates (z-values given below estimates). Older respondents tend to be more afraid than younger respondents, in particular concerning terrorism but less concerning climate change. Females have for all items higher fear levels than males, the effects of EastWest are rather mixed, people from the Eastern parts of the country are more afraid of globalization but less afraid of nuclear energy. Higher education seems to reduce the level of fears. The necessity of covariates is also supported by testing. The log-likelihood test that compares the model without covariates to the model with covariates is 88.081on 10 df. Thus, the covariates turn out to be influential if one accounts for the heterogeneity in the population.

Table 11.1 also shows the effects of covariates if they are assumed to be global, that is, not item-specific. It is seen that the model with item-specific covariate effects

Table 11.2 Testing if covariates can considered global, one at a time

	Age	Gender	EastWest	Abitur	Log-likelihood	LR test	df
	Var	Var	Var	Var	−1684.871		
	Global	Var	Var	Var	−1695.435	21.128	3
	Var	Global	Var	Var	−1708.562	47.382	3
	Var	Var	Global	Var	−1704.869	39.996	3
	Var	Var	Var	Global	−1702.910	36.079	3

fits much better. However, it is unclear which variables have item-specific effects and which ones could be considered as global. Therefore, in Table 11.2 the log-likelihoods are given if one of the variables is considered as having global effects, and the corresponding likelihood ratio tests are computed. It is seen that none of the variables should be considered as global. □

Variable Selection by Penalization

Penalization methods as already considered in Chap. 10 can be useful to select which variables are relevant and in what way. They are based on penalized log-likelihoods instead of simple likelihood maximization. The usual (marginal) log-likelihood is replaced by

$$l_p(\{\gamma_i\}, \{\delta_{ir}\}) = l(\{\gamma_i\}, \{\delta_{ir}\}) - P_{\boldsymbol{\lambda}}(\{\gamma_i\}),$$

where $l(.)$ is the usual log-likelihood and $P_{\boldsymbol{\lambda}}(\{\gamma_i\})$ is a penalty term depending on a possibly multidimensional tuning parameter $\boldsymbol{\lambda}$. With selection of effects in mind it is useful to consider the vectors $\boldsymbol{\gamma}_{j,\text{dif}}^T = (\gamma_{1j} - \gamma_{2j}, \gamma_{2j} - \gamma_{3j}, \dots)$, which contain all differences of effects between pairs of items for the jth variable. A penalty term with a two-dimensional tuning parameter is

$$P_{\boldsymbol{\lambda}}(\{\gamma_i\}) = \lambda_f \sum_{j=1}^{m} ||\boldsymbol{\gamma}_{j,\text{dif}}|| + \lambda_s ||\boldsymbol{\gamma}_{.j}||,$$

where $||.||$ is the L_2-norm and $\boldsymbol{\gamma}_{.j}^T = (\gamma_{1j} \dots, \gamma_{Ij})$ is the vector that collects all parameters linked to the jth variable. The first term is a fusion term that enforces that effects are global. For $\lambda_f \to \infty$ one obtains $\gamma_{1j} = \cdots = \gamma_{Ij}$, that means the effect of the jth variable is the same for all items. The second term is a selection term, for $\lambda_s \to \infty$ all parameters become zero, $\gamma_{1j} = \cdots = \gamma_{Ij} = 0$. Thus, the penalty allows to evaluate which variables are global and which variables can be totally omitted. The construction uses the grouped lasso methodology (Yuan and Lin, 2006) for single penalty terms but combines them to enforce differing effects. Various combinations of penalties that enforce selection and shrinkage to global effects have been considered, for example, by Tutz and Gertheiss (2016), Oelker and Tutz (2017), see also the composite absolute penalty family proposed by Zhao et al. (2009).

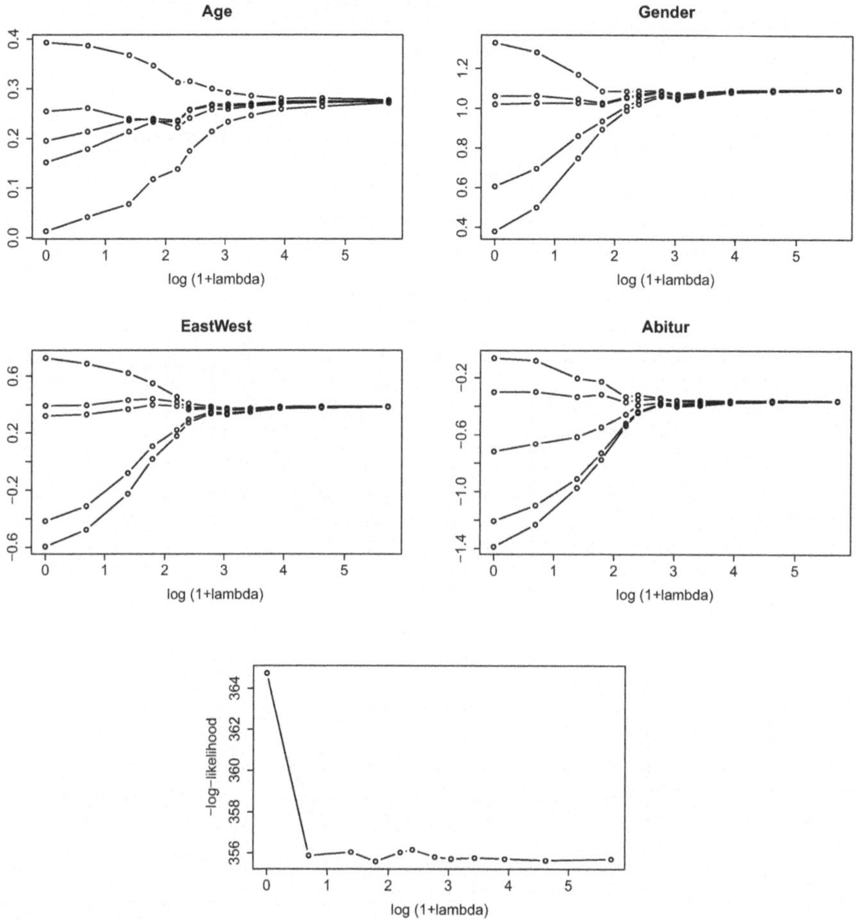

Fig. 11.3 Coefficient paths for grouped lasso on differences; last panel is cross validation loss

For illustration Fig. 11.3 shows the coefficients as a function of the tuning parameter if only the first term in the penalty is used to distinguish between item-specific and global parameters ($\lambda_s = 0$). Figure 11.4 shows the coefficients if both penalty terms are used ($\lambda_f = \lambda_s$). It also shows the negative log-likelihood (Kullback-Leibler loss) for five-fold cross validation, which serves to determine which tuning parameter should be chosen. It is minimal if $\log(1 + \lambda)$ in the range between 2 and 3. In this range parameters are not yet estimated as global or zero.

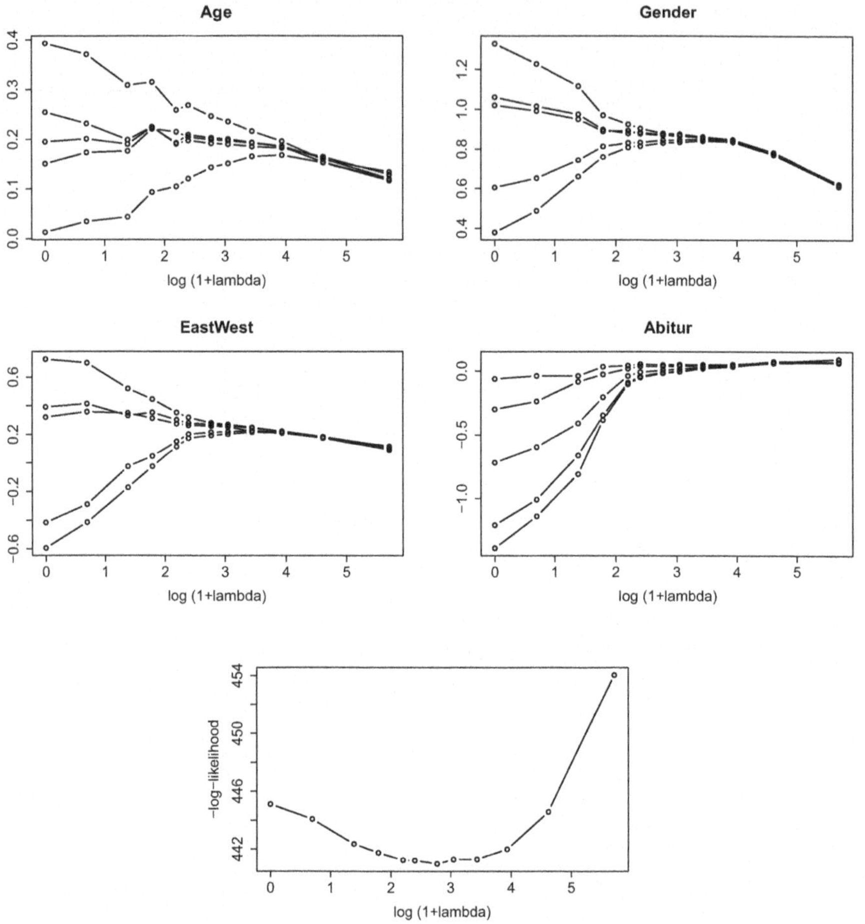

Fig. 11.4 Age, gender, eastwest, abitur for grouped lasso on differences and selection; last panel is cross validation loss

11.3 Including Item-Specific Covariates

In the following the inclusion of subject and item-specific covariates is considered for binary response models as propagated by De Boeck and Wilson (2004, 2016).

Let $\boldsymbol{x}_p$ denote a person-specific vector of covariates and $\boldsymbol{z}_i$ an item-specific parameters. The predictor η_{pi} in binary response models of the form $P(Y_{pi} = 1) = F(\eta_{pi})$ without covariates is given by $\eta_{pi} = \theta_p - \delta_i$. A general form of the predictor that includes covariates is

$$\eta_{pi} = \theta_p + \boldsymbol{z}_{pi}^T \boldsymbol{\beta} - \delta_i,$$

Table 11.3 Linear predictor and specifications of effects

$z_{pi}^T\boldsymbol{\beta}$	General structure
$\boldsymbol{x}_p^T\gamma + z_i^T\xi$	Global effects of subject-specific covariates
	$z_{pi}^T = (\boldsymbol{x}_p^T, z_i^T), \boldsymbol{\beta}^T = (\gamma^T, \xi^T)$
$\boldsymbol{x}_p^T\gamma_i + z_i^T\xi$	Item-specific effects of subject-specific covariates
	$z_{pi}^T = (\mathbf{0}^T, \ldots \boldsymbol{x}_p^T, \ldots \mathbf{0}^T, z_i^T), \boldsymbol{\beta}^T = (\gamma_1^T, \ldots, \gamma_I^T, \xi^T)$
	Where $\mathbf{0}$ is a vector of zeros

where z_{pi} is constructed from $\boldsymbol{x}_p$ and z_i. It includes the case of global effects of person-specific covariates $\boldsymbol{x}_p$ and item-specific covariates z_i,

$$\eta_{pi} = \theta_p + \boldsymbol{x}_p^T\gamma + z_i^T\xi - \delta_i \tag{11.8}$$

with $z_{pi}^T = (\boldsymbol{x}_p^T, z_i^T), \boldsymbol{\beta}^T = (\gamma^T, \xi^T)$. Also item-specific effects of subject-specific covariates can be modeled in the form

$$\eta_{pi} = \theta_p + \boldsymbol{x}_p^T\gamma_i + z_i^T\xi - \delta_i,$$

by specifying $z_{pi}^T = (\mathbf{0}^T, \ldots, \boldsymbol{x}_p^T, \ldots \mathbf{0}^T, z_i^T)$, which is an extended vector that contains vectors of zeros and $\boldsymbol{\beta}^T = (\gamma_1^T, \ldots, \gamma_I^T, \xi^T)$ collects all the weights (Table 11.3).

Example 11.5 Verbal aggression
As an illustrating example we consider again the verbal aggression data VerbAgg from package *lm4*, which have also been used in Example 6.4. Various models with explanatory variables have been considered by De Boeck and Wilson (2004). We use only some of the variables in the data set. Each item consists of three verbally aggressive behaviors (cursing, scolding,wanting), distinguishes between self-to-blame and others-to blame, and two behavioral modes (wanting, doing).

Subject-specific variables are: Anger (the subject's trait anger score as measured on the State-Trait Anger Expression Inventory (STAXI)) and Gender (1: male, 0: female). Item-specific variables are: mode (behavior mode—a factor with levels want and do), situ (situation type—a factor with levels other and self indicating other-to-blame and self-to-blame) and btype behavior type—afactor with levels curse, scold, and shout).

Responses are in three categories or binary: resp (the subject's response to the item—an ordered factor with levels no, perhaps, yes), r2 (dichotomous version of the response—a factor with levels 1: perhaps or yes 0: no.)

The models can be fitted by using the function *glmer*. With only subject-specific covariates and assuming that person and items effects are random one can use for binary responses:

```
library("lme4")
print(m1 <- glmer(r2 ~ Anger + Gender + (1|id) + (1|item),
   family = binomial,data = VerbAgg, nAGQ=0L), corr=TRUE)
summary(m1)
```

The resulting output shows that the anger score increases the probability of a response in category 1 significantly while gender shows no significant effect.

```
Generalized linear mixed model fit by maximum likelihood
 (Adaptive Gauss-Hermite Quadrature, nAGQ = 0) ['glmerMod']
 Family: binomial  ( logit )
Formula: r2 ~ Anger + Gender + (1 | id) + (1 | item)
   Data: VerbAgg

     AIC      BIC   logLik deviance df.resid
  8199.2   8233.8  -4094.6   8189.2     7579

Scaled residuals:
    Min      1Q  Median      3Q     Max
-5.9861 -0.6244 -0.1787  0.6267 14.0150

Random effects:
 Groups Name        Variance Std.Dev.
 id     (Intercept) 1.797    1.341
 item   (Intercept) 1.275    1.129
Number of obs: 7584, groups:  id, 316; item, 24
```

```
Fixed effects:
            Estimate Std. Error z value Pr(>|z|)
(Intercept) -1.31362    0.41811  -3.142  0.00168 **
Anger        0.05440    0.01676   3.245  0.00118 **
GenderM      0.30392    0.19135   1.588  0.11222

Correlation of Fixed Effects:
        (Intr) Anger
Anger   -0.805
GenderM -0.124  0.022
```

Fitting with subject and item-specific covariates can be obtained by

```
print(m3 <- glmer(r2 ~ Anger + Gender  + mode+situ+ btype+
   (1|id) + (1|item), family = binomial,
  data = VerbAgg, nAGQ=0L), corr=TRUE)
summary(m3)
```

yielding

```
Generalized linear mixed model fit by maximum likelihood
(Adaptive Gauss-Hermite Quadrature, nAGQ = 0) ['glmerMod']
 Family: binomial  ( logit )
Formula: r2 ~ Anger + Gender + mode +
         situ + btype + (1 | id) + (1 | item)
   Data: VerbAgg

     AIC      BIC   logLik deviance df.resid
  8154.2   8216.6  -4068.1   8136.2     7575

Scaled residuals:
    Min      1Q  Median      3Q     Max
-6.2557 -0.6240 -0.1777  0.6247 13.1433

Random effects:
 Groups Name        Variance Std.Dev.
 id     (Intercept) 1.7923   1.3387
 item   (Intercept) 0.1188   0.3446
Number of obs: 7584, groups:  id, 316; item, 24

Fixed effects:
            Estimate Std. Error z value Pr(>|z|)
(Intercept)  0.54854    0.38564   1.422  0.15490
Anger        0.05438    0.01674   3.248  0.00116
GenderM      0.30424    0.19112   1.592  0.11141
modedo      -0.67909    0.15186  -4.472 7.75e-06
situself    -1.01253    0.15204  -6.660 2.74e-11
btypescold  -1.01747    0.18521  -5.493 3.94e-08
btypeshout  -2.02064    0.18752 -10.776  < 2e-16
```

It is seen that the effects of anger and gender hardly change when item-specific variables are included. Items that refer to doing tend to yield less often verbally aggressive behaviour than items than refer to wanting. Also self-blaming items have weaker responses than items that refer to blaming others. Finally scolding and shouting items yield weaker responses than cursing items. □

Binary models with predictor (11.8) can also be applied to polytomous responses if one assumes a hierarchical structure of responses as in tree-based item response models (Chap. 9), see De Boeck and Wilson (2004, 2016). The authors also consider more general, multidimensional models that include explanatory variables.

11.4 Fitting of Models

The R package MultOrdRS provides methods to explain the effects of covariates on the response style as proposed by Schauberger and Tutz (2022). The method is implemented both for the cumulative model (graded response model) and the adjacent categories model.

The R package *lme4* fits linear and generalized linear mixed-effects models. It contains the useful function *glmer*.

The R package *hIRT* can be used to fit the cumulative version of Zhou's hierarchical item response model (Zhou, 2019), which assumes that the mean and variance of the one-dimensional latent trait may depend on covariates.

Chapter 12
R Packages

R is a language and environment for statistical computing and graphics (R Core Team, 2021). It is available as free software under the terms of the Free Software Foundation's GNU General Public License and provides a wide variety of statistical and graphical techniques. The main uniform resource locator of R is http://www.r-project.org.

Packages are collections of R functions, data, and compiled code in a well-defined format, created to add specific functionality. For a current and continuously updated overview of general psychometric models and methods available in R see

https://cran.r-project.org/view=Psychometrics, (Mair et al., 2023).

A detailed description of various R packages for item response theory analysis is found in Choi and Asilkalkan (2019). In the following only a small selection of R packages is briefly described.

Binary and Polytomous Models

- R package *ltm* (Rizopoulos, 2006): Analysis of multivariate dichotomous and polytomous responses, includes the Rasch, the Two-Parameter Logistic, the Birnbaum's Three-Parameter, the Graded Response, and the Generalized Partial Credit Models; yields marginal estimates.
- R package *TAM* (Robitzsch et al., 2024): Includes marginal maximum likelihood estimation and joint maximum likelihood estimation for unidimensional and multidimensional item response models. The package functionality covers the Rasch model, 2PL model, 3PL model, generalized partial credit model, multi-faceted Rasch model, nominal item response model, structured latent class model, mixture distribution IRT models, and located latent class models. Latent regression models and plausible value imputation are also supported.

G. Tutz, *A Short Guide to Item Response Theory Models*, Statistics for Social and Behavioral Sciences, https://doi.org/10.1007/978-3-031-87271-6_12

- Package *eRm* (Mair and Hatzinger, 2007b; Mair et al., 2016): Provides conditional fitting of Rasch models, linear logistic test models, the rating scale model, linear rating scale models, partial credit models, and linear partial credit models (LPCM). Additional features are the ML estimation of the person parameters, Andersen's LR test, item-specific Wald test, Martin-Loef-test.

Extended Polytomous Models

- R package UPCM: An implementation of PCMRS (Partial Credit Model with Response Styles) as proposed by Tutz et al. (2018). PCMRS is an extension of the regular partial credit model. PCMRS allows for an additional person parameter that characterizes the response style of the person.
- R package MultOrdRS: Provides methods to account for personal response styles and to explain the effects of covariates on the response style, as proposed Schauberger and Tutz (2022). The method is implemented both for the cumulative model (graded response model) and the adjacent categories model.

Count Data Models

- R package *glmmTMB*: Fits the Poisson Rasch model and the Conway-Maxwell model. (Forthmann et al., 2020).

Continuous Responses

- Package *pcIRT* (Hohensinn, 2018): provides functions for estimating IRT models for polytomous (nominal) and continuous data.

Differential Item Functioning

- R package *difR*: A comprehensive package for the investigation of DIF linked to group membership, includes many methods, among them, generalized Lord, generalized Mantel-Haenszel, logistic regression, Lord's chi-square test, Mantel-Haenszel, Raju's area, see Magis et al. (2010).
- R package *DIFlasso*: Penalized Ml estimation for Rasch models, allows the simultaneous selection of items and variables that induce differential item functioning (Tutz and Schauberger, 2015).

- R package *GPCMlasso*: Allows to identify DIF in Generalized Partial Credit Models. Several variables can be considered simultaneously. For each pair between variable and item, a parametric DIF effect is introduced which indicates DIF if the respective parameter is selected (Schauberger and Mair, 2020).
- R package *DIFtree*: Item focussed recursive partitioning for simultaneous selection of items and variables that induce differential item functioning (DIF) in dichotomous or polytomous items (Tutz and Berger, 2016a; Berger and Tutz, 2016).
- R package *psychotree*: Rasch trees as proposed by Strobl et al. (2015).

Tree-Based Models

- R package *irtrees*: Fits tree-based item response models of the generalized mixed model family, in particular binary trees as propagated by De Boeck and Partchev (2012).

Explanatory Item Response Models

- R package MultOrdRS: Provides methods to explain the effects of covariates on the response style, as proposed Schauberger and Tutz (2022). The method is implemented both for the cumulative model (graded response model) and the adjacent categories model.
- R package *hIRT*: Fits the cumulative version of Zhou's hierarchical item response model (Zhou, 2019), which assumes that the mean and variance of the one-dimensional latent trait may depend on covariates.
- R package *lme4* (Bates et al., 2015): Fits linear and generalized linear mixed-effects models. In particular the function *glmer* is helpful.

Chapter 13
Examples

Chapter 2 Binary Rasch models

- FIMS study, Example 2.1
- Independence, Example 2.2
- FIMS data, Example 2.3
- Fears data, Example 2.4
- FIMS study, Example 2.5

Chapter 3 Alternative binary models

- Fears data, Example 3.1
- Knowledge Data, Example 3.2
- FIMS study, Example 3.3

Chapter 4 Ordinal Models

- FIMS study, Example 4.1
- FIMS study, Example 4.2
- FIMS study, Example 4.3
- FIMS study, Example 4.4
- Fears Data, Example 4.5

Chapter 5 Extended Models

- Science data, Example 5.1
- Tenseness, Example 5.2
- Tenseness, Example 5.3

G. Tutz, *A Short Guide to Item Response Theory Models*, Statistics for Social and Behavioral Sciences, https://doi.org/10.1007/978-3-031-87271-6_13

Chapter 6 Thresholds

- Self-regulation, Chap. 6, Example 6.1
- Fears data, Chap. 6, Example 6.2,
- Fears data, Chap. 6, Example 6.3,
- Verbal aggression, Chap. 6, Example 6.4
- Decathlon data, Chap. 6, Example 6.5

Chapter 8 Count

- Verbal Fluency, Chap. 8, Example 8.1,
- Verbal Fluency, Chap. 8, Example 8.2,
- Verbal Fluency, Chap. 8, Example 8.3

Chapter 9 Trees

- Verbal aggression, Chaps. 9, Sect. 9.1

Chapter 10 Differential Item Functioning

- Knowledge Data, Example 10.1,
- Fears Data, Example 10.2,
- Knowledge Data, Example 10.3,
- Knowledge Data, Example 10.4,
- CTB Science data, Example 10.5
- Sleep Study 7.1

Chapter 11 Explanatory variables

- Fears data, Example 11.1
- Fears data, Example 11.3
- Fears data, Example 11.2
- Fears data, Example 11.4
- Verbal aggression, Example 11.5

Bibliography

Adams, R. J., M. L. Wu, and M. Wilson (2012). The Rasch rating model and the disordered threshold controversy. *Educational and Psychological Measurement 72*(4), 547–573.

Agresti, A. (2013). *Categorical Data Analysis, 3d Edition*. New York: Wiley.

Akaike, H. (1973). In B. Petrov and F. Caski (Eds.), *Information Theory and the Extension of the Maximum Likelihood Principle*, Second International Symposium on Information Theory. Akademia Kiado.

Andersen, E. and M. Madsen (1977). Estimating the parameters of the latent population distribution. *Psychometrika 42*(3), 357–374.

Andersen, E. B. (1973). A goodness of fit test for the Rasch model. *Psychometrika 38*(1), 123–140.

Andersen, E. B. (1977). Sufficient statistics and latent trait models. *Psychometrika 42*, 69–81.

Anderson, D. A. and M. Aitkin (1985). Variance component models with binary response: Interviewer variability. *Journal of the Royal Statistical Society Series B 47*, 203–210.

Andrich, D. (1982). An extension of the Rasch model for ratings providing both location and dispersion parameters. *Psychometrika 47*(1), 105–113.

Andrich, D. (1988). The application of an unfolding model of the pirt type to the measurement of attitude. *Applied psychological measurement 12*(1), 33–51.

Andrich, D. (1996). A hyperbolic cosine latent trait model for unfolding polytomous responses: Reconciling thurstone and Likert methodologies. *British Journal of Mathematical and Statistical Psychology 49*(2), 347–365.

Andrich, D. (2010). Sufficiency and conditional estimation of person parameters in the polytomous Rasch model. *Psychometrika 75*(2), 292–308.

Andrich, D. (2013). An expanded derivation of the threshold structure of the polytomous Rasch model that dispels any 'threshold disorder controversy'. *Educational and Psychological Measurement 73*(1), 78–124.

Andrich, D. (2015). The problem with the step metaphor for polytomous models for ordinal assessments. *Educational Measurement: Issues and Practice 34*(2), 8–14.

Andrich, D. (2016). Rasch rating-scale model. In W. Van der Linden (Ed.), *Handbook of Modern Item Response Theory*, pp. 75–94. Chapman and Hall/CRC.

Andrich, D. and G. Luo (1993). A hyperbolic cosine latent trait model for unfolding dichotomous single-stimulus responses. *Applied Psychological Measurement 17*(3), 253–276.

Andrich, D. and I. Marais (2019). *A course in Rasch measurement theory - Measuring in the educational, social and health sciences*. Springer.

Baker, F. B., S.-H. Kim, et al. (2017). *The basics of item response theory using R*. Springer.

G. Tutz, *A Short Guide to Item Response Theory Models*, Statistics for Social and Behavioral Sciences, https://doi.org/10.1007/978-3-031-87271-6

Bates, D., M. Mächler, B. Bolker, and S. Walker (2015). Fitting linear mixed-effects models using lme4. *Journal of Statistical Software 67*(1), 1–48.

Battauz, M. and R. Bellio (2021). Shrinkage estimation of the three-parameter logistic model. *British Journal of Mathematical and Statistical Psychology 74*(3), 591–609.

Baumgartner, H. and J.-B. E. Steenkamp (2001). Response styles in marketing research: A cross-national investigation. *Journal of Marketing Research 38*(2), 143–156.

Bazán, J. L., H. Bolfarine, and M. D. Branco (2006). A skew item response model. *Bayesian analysis 1*(4), 861–892.

Berger, M. (2020). *DIFtree: Item Focussed Trees for the Identification of Items in Differential Item Functioning*. R package version 3.1.6.

Berger, M. and G. Tutz (2016). Detection of uniform and non-uniform differential item functioning by item focussed trees. *Journal of Educational and Behavioral Statistics 40*, 559–592.

Birnbaum, A. (1968). Some latent trait models and their use in inferring an examinee's ability. In F. M. Lord and M. R. Novick (Eds.), *Statistical theories of mental test scores*, pp. 392–479. Addison-Wesley.

Bock, R. D. (1972). Estimating item parameters and latent ability when responses are scored in two or more nominal categories. *Psychometrika 37*(1), 29–51.

Böckenholt, U. (2012). Modeling multiple response processes in judgment and choice. *Psychological Methods 17*(4), 665–678.

Böckenholt, U. (2017). Measuring response styles in Likert items. *Psychological methods* (22), 69–83.

Böckenholt, U. and T. Meiser (2017). Response style analysis with threshold and multi-process irt models: A review and tutorial. *British Journal of Mathematical and Statistical Psychology 70*(1), 159–181.

Bollmann, S., M. Berger, and G. Tutz (2018). Item-focused trees for the detection of differential item functioning in partial credit models. *Educational and Psychological Measurement 78*(5), 781–804.

Bolt, D. M. and T. R. Johnson (2009). Addressing score bias and differential item functioning due to individual differences in response style. *Applied Psychological Measurement 33*(5), 335–352.

Bolt, D. M. and X. Liao (2022). Item complexity: A neglected psychometric feature of test items? *Psychometrika*, 1–19.

Bolt, D. M. and J. R. Newton (2011). Multiscale measurement of extreme response style. *Educational and Psychological Measurement 71*(5), 814–833.

Braun, H. I. and R. Mislevy (2005). Intuitive test theory. *Phi Delta Kappan 86*(7), 488–497.

Breiman, L., J. H. Friedman, R. A. Olshen, and J. C. Stone (1984). *Classification and Regression Trees*. Monterey, CA: Wadsworth.

Bühlmann, P. and S. Van De Geer (2011). *Statistics for High-Dimensional Data: Methods, Theory and Applications*. Springer-Verlag New York.

Carifio, J. and R. J. Perla (2007). Ten common misunderstandings, misconceptions, persistent myths and urban legends about Likert scales and Likert response formats and their antidotes. *Journal of Social Sciences 3*(3), 106–116.

Chen, Y., C. Li, J. Ouyang, and G. Xu (2023). DIF statistical inference without knowing anchoring items. *Psychometrika 88*(4), 1097–1122.

Choi, Y.-J. and A. Asilkalkan (2019). R packages for item response theory analysis: Descriptions and features. *Measurement: Interdisciplinary research and perspectives 17*(3), 168–175.

Christensen, K. B., J. B. Bjorner, S. Kreiner, and J. H. Petersen (2002). Testing unidimensionality in polytomous Rasch models. *Psychometrika 67*, 563–574.

Coombs, C. H. (1964). *A theory of data*. Wiley.

De Boeck, P. and I. Partchev (2012). Irtrees: Tree-based item response models of the glmm family. *Journal of Statistical Software 48*(1), 1–28.

De Boeck, P. and I. Partchev (2012). Irtrees: Tree-based item response models of the glmm family. *Journal of Statistical Software – Code Snippets 48*(1), 1–28.

De Boeck, P. and M. Wilson (2004). *Explanatory item response models: A generalized linear and nonlinear approach.* Springer Verlag.

De Boeck, P. and M. R. Wilson (2016). Explanatory response models. In W. Van der Linden (Ed.), *Handbook of Item Response Theory, Volume One*, pp. 593–608. Chapman and Hall/CRC.

De Leeuw, J. and N. Verhelst (1986). Maximum likelihood estimation in generalized Rasch models. *Journal of educational statistics 11*(3), 183–196.

Debelak, R., C. Strobl, and M. D. Zeigenfuse (2022). *An introduction to the Rasch model with examples in R.* Chapman and Hall/CRC.

Deng, S. and D. M. Bolt (2016). A sequential irt model for multiple-choice items and a multidimensional extension. *Applied Psychological Measurement 40*(4), 243–257.

Dierckx, P. (1993). *Curve and Surface Fitting with Splines.* Oxford: Oxford Science Publications.

Dingle, H. (1950). A theory of measurement. *The British Journal for the Philosophy of Science 1*(1), 5–26.

Doebler, A. and H. Holling (2016). A processing speed test based on rule-based item generation: An analysis with the Rasch poisson counts model. *Learning and individual differences 52*, 121–128.

Drasgow, F., O. S. Chernyshenko, and S. Stark (2010). 75 years after Likert: Thurstone was right! *Industrial and Organizational Psychology 3*(4), 465–476.

Draxler, C. and A. Kurz (2025). Testing measurement invariance in a conditional likelihood framework by considering multiple covariates simultaneously. *Behavior Research Methods 57*(1), 50.

Draxler, C., A. Kurz, C. Gürer, and J. P. Nolte (2024). An improved inferential procedure to evaluate item discriminations in a conditional maximum likelihood framework. *Journal of Educational and Behavioral Statistics*, 403–430.

Draxler, C., G. Tutz, K. Zink, and C. Gürer (2016). Comparison of maximum likelihood with conditional pairwise likelihood estimation of person parameters in the Rasch model. *Communications in Statistics-Simulation and Computation 45*(6), 2007–2017.

Edwards, A. L. and L. L. Thurstone (1952). An internal consistency check for scale values determined by the method of successive intervals. *Psychometrika 17*, 169–180.

Eggen, T. J. (2000). On the loss of information in conditional maximum likelihood estimation of item parameters. *Psychometrika 65*(3), 337–362.

Eggen, T. J. and N. D. Verhelst (2006). Loss of information in estimating item parameters in incomplete designs. *Psychometrika 71*, 303–322.

Eid, M. and M. Rauber (2000). Detecting measurement invariance in organizational surveys. *European Journal of Psychological Assessment 16*(1), 20–30.

Eilers, P. H. and B. D. Marx (2021). *Practical Smoothing: The Joys of P-splines.* Cambridge University Press.

Eilers, P. H. C. and B. D. Marx (1996). Flexible smoothing with B-splines and Penalties. *Statistical Science 11*, 89–121.

Fahrenberg, J. (2010). Freiburg Complaint Checklist [Freiburger Beschwerdenliste (FBL)]. *Göttingen, Hogrefe.*

Falk, C. F. and L. Cai (2016). A flexible full-information approach to the modeling of response styles. *Psychological methods 21*(3), 328.

Fischer, G. H. (1981). On the existence and uniqueness of maximum-likelihood estimates in the Rasch model. *Psychometrika 46*(1), 59–77.

Fischer, G. H. (1995). Derivations of the Rasch model. In G. Fischer and I. Molenaar (Eds.), *Rasch models*, pp. 15–38. Springer.

Fjørtoft, I., A. V. Pedersen, H. Sigmundsson, and B. Vereijken (2011). Measuring physical fitness in children who are 5 to 12 years old with a test battery that is functional and easy to administer. *Physical therapy 91*(7), 1087–1095.

Follmann, D. (1988). Consistent estimation in the Rasch model based on nonparametric margins. *Psychometrika 53*(4), 553–562.

Forthmann, B. and P. Doebler (2021). Reliability of researcher capacity estimates and count data dispersion: a comparison of poisson, negative binomial, and conway-maxwell-poisson models. *Scientometrics 126*(4), 3337–3354.

Forthmann, B., D. Gühne, and P. Doebler (2020). Revisiting dispersion in count data item response theory models: The Conway–Maxwell–Poisson counts model. *British Journal of Mathematical and Statistical Psychology 73*(1), 32–50.

Forthmann, B., H. Holling, P. Çelik, M. Storme, and T. Lubart (2017). Typing speed as a confounding variable and the measurement of quality in divergent thinking. *Creativity Research Journal 29*(3), 257–269.

Fox, J.-P. (2010). *Bayesian item response modeling: Theory and applications*. Springer.

Glas, C. A. (2007). Testing generalized Rasch models. In M. von Davier and C. Carstensen (Eds.), *Multivariate and mixture distribution Rasch models*, pp. 37–55. Springer.

Glas, C. A. and N. D. Verhelst (1995). Testing the Rasch model. In G. H. Fischer and I. W. Molenaar (Eds.), *Rasch models: Foundations, recent developments, and applications*, pp. 69–95. Springer.

Goldstein, H. (1987). *Multilevel Models in Educational and Social Research*. New York: London and Oxford University Press.

Gollwitzer, M., M. Eid, and R. Jürgensen (2005). Response styles in the assessment of anger expression. *Psychological assessment 17*(1), 56.

Gürer, C. and C. Draxler (2023). Penalization approaches in the conditional maximum likelihood and rasch modelling context. *British Journal of Mathematical and Statistical Psychology 76*(1), 154–191.

Haberman, S. J. (1977). Maximum likelihood estimates in exponential response models. *Annals of Statistics 5*, 815–841.

Hand, D. J. (1996). Statistics and the theory of measurement. *Journal of the Royal Statistical Society: Series A (Statistics in Society) 159*(3), 445–473.

Hanson, B. A. and A. A. Béguin (2002). Obtaining a common scale for item response theory item parameters using separate versus concurrent estimation in the common-item equating design. *Applied psychological measurement 26*(1), 3–24.

Harpe, S. E. (2015). How to analyze Likert and other rating scale data. *Currents in pharmacy teaching and learning 7*(6), 836–850.

Hastie, T. and R. Tibshirani (1986). Generalized additive models (c/r: p. 310–318). *Statist. Sci. 1*, 297–310.

Heller, G. Z., M. Manuguerra, and R. Chow (2016). How to analyze the visual analogue scale: Myths, truths and clinical relevance. *Scandinavian journal of pain 13*(1), 67–75.

Hemker, B. T., K. Sijtsma, I. W. Molenaar, and B. W. Junker (1997). Stochastic ordering using the latent trait and the sum score in polytomous irt models. *Psychometrika 62*(3), 331–347.

Hemker, B. T., L. A. van der Ark, and K. Sijtsma (2001). On measurement properties of continuation ratio models. *Psychometrika 66*(4), 487–506.

Hilbe, J. (2011). *Negative binomial regression*. Cambridge University Press.

Hinde, J. (1982). Compound poisson regression models. In R. Gilchrist (Ed.), *GLIM 1982 International Conference on Generalized Linear Models*, pp. 109–121. New York: Springer-Verlag.

Hoerl, A. E. and R. W. Kennard (1970). Ridge regression: Bias estimation for nonorthogonal problems. *Technometrics 12*, 55–67.

Hohensinn, C. (2018). pcirt: an R package for polytomous and continuous Rasch models. *Journal of Statistical Software 84*, 1–14.

Hoijtink, H. and A. Boomsma (1995). On person parameter estimation in the dichotomous Rasch model. In G. Fischer and I. Molenaar (Eds.), *Rasch model models: Foundations, recent developments, and applications*, pp. 53–68. Springer.

Holland, P. W. and M. Hoskens (2003). Classical test theory as a first-order item response theory: Application to true-score prediction from a possibly nonparallel test. *Psychometrika 68*(1), 123–149.

Holland, P. W. and P. R. Rosenbaum (1986). Conditional association and unidimensionality in monotone latent variable models. *The Annals of Statistics*, 1523–1543.

Holland, P. W. and D. T. Thayer (1985). An alternate definition of the ets delta scale of item difficulty. *ETS Research Report Series 1985*(2), i–10.

Holland, P. W. and D. T. Thayer (1988). Differential item performance and the Mantel-Haenszel procedure. *Test validity*, 129–145.

Holland, W. and W. H. (1993). *Differential Item Functioning*. Lawrence Erlbaum Associates.

Hopkins, W. G., E. J. Schabort, and J. A. Hawley (2001). Reliability of power in physical performance tests. *Sports medicine 31*, 211–234.

Hothorn, T., K. Hornik, and A. Zeileis (2006). Unbiased recursive partitioning: A conditional inference framework. *Journal of Computational and Graphical Statistics 15*, 651–674.

Huang, A. (2017). Mean-parametrized conway–maxwell–poisson regression models for dispersed counts. *Statistical Modelling 17*(6), 359–380.

Hung, L.-F. (2012). A negative binomial regression model for accuracy tests. *Applied Psychological Measurement 36*(2), 88–103.

Irtel, H. (1995). An extension of the concept of specific objectivity. *Psychometrika 60*(1), 115–118.

Jackson, P. H. and C. C. Agunwamba (1977). Lower bounds for the reliability of the total score on a test composed of non-homogeneous items: I: Algebraic lower bounds. *Psychometrika 42*(4), 567–578.

Jamieson, S. (2004). Likert scales: how to (ab) use them. *Medical education 38*(12), 1217–1218.

Jansen, M. G. (1995). The Rasch poisson counts model for incomplete data: An application of the em algorithm. *Applied Psychological Measurement 19*(3), 291–302.

Jansen, P. G. and E. E. Roskam (1986). Latent trait models and dichotomization of graded responses. *Psychometrika 51*(1), 69–91.

Jeon, M. and P. De Boeck (2016). A generalized item response tree model for psychological assessments. *Behavior research methods 48*(3), 1070–1085.

Johnson, M. S. (2007). Marginal maximum likelihood estimation of item response models in r. *Journal of Statistical Software 20*, 1–24.

Johnson, T. R. (2003). On the use of heterogeneous thresholds ordinal regression models to account for individual differences in response style. *Psychometrika 68*(4), 563–583.

Johnson, T. R. and D. M. Bolt (2010). On the use of factor-analytic multinomial logit item response models to account for individual differences in response style. *Journal of Educational and Behavioral Statistics 35*(1), 92–114.

Kaptein, M. C., C. Nass, and P. Markopoulos (2010). Powerful and consistent analysis of Likert-type ratingscales. In *Proceedings of the SIGCHI Conference on Human Factors in Computing Systems*, pp. 2391–2394. ACM.

Kelderman, H. (1996). Multidimensional Rasch models for partial-credit scoring. *Applied Psychological Measurement 20*(2), 155–168.

Khorramdel, L. and M. von Davier (2014). Measuring response styles across the big five: A multiscale extension of an approach using multinomial processing trees. *Multivariate Behavioral Research 49*(2), 161–177.

Kim, J.-S. and B. A. Hanson (2002). Test equating under the multiple-choice model. *Applied Psychological Measurement 26*(3), 255–270.

Kim, S. and W.-C. Lee (2006). An extension of four irt linking methods for mixed-format tests. *Journal of Educational Measurement 43*(1), 53–76.

Kim, S.-H., A. S. Cohen, and T.-H. Park (1995). Detection of differential item functioning in multiple groups. *Journal of Educational Measurement 32*(3), 261–276.

Kolen, M. J. and R. L. Brennan (2014). *Test equating, scaling, and linking: Methods and practices*. Springer Science & Business Media.

Krantz, D. H., R. D. Luce, P. Suppes, and A. Tversky (1971). *Foundations of Measurement*, Volume 1. New York: Academic Press.

Kulas, J. T., A. A. Stachowski, and B. A. Haynes (2008). Middle response functioning in Likert-responses to personality items. *Journal of Business and Psychology 22*(3), 251–259.

Lakes, K. D. and W. T. Hoyt (2009). Applications of generalizability theory to clinical child and adolescent psychology research? *Journal of Clinical Child & Adolescent Psychology 38*(1), 144–165.

Lantz, B. (2013). Equidistance of Likert-type scales and validation of inferential methods using experiments and simulations. *The Electronic Journal of Business Research Methods 11*(1), 16–28.

Likert, R. (1932). A technique for the measurement of attitudes. *Archives of psychology* (140), 44–53.

Lin, Y., A. Brown, and P. Williams (2023). Multidimensional forced-choice cat with dominance items: an empirical comparison with optimal static testing under different desirability matching. *Educational and Psychological Measurement 83*(2), 322–350.

Lord, F. (1952). A theory of test scores. *Psychometric monographs No. 7.*

Lord, F. M. (1980). *Applications of item response theory to practical testing problems.* Routledge.

Lord, F. M. and M. R. Novick (1968). *Statistical theories of mental test scores.* Addison-Wesley Publishing Company.

Luce, R. D., P. Suppes, and D. H. Krantz (2007). *Foundations of measurement: representation, axiomatization, and invariance.* Courier Corporation.

Magis, D. (2013). A note on the item information function of the four-parameter logistic model. *Applied Psychological Measurement 37*(4), 304–315.

Magis, D., S. Bèland, F. Tuerlinckx, and P. Boeck (2010). A general framework and an R package for the detection of dichotomous differential item functioning. *Behavior Research Methods 42*(3), 847–862.

Magis, D., G. Raîche, S. Béland, and P. Gérard (2011). A generalized logistic regression procedure to detect differential item functioning among multiple groups. *International Journal of Testing 11*(4), 365–386.

Magis, D., F. Tuerlinckx, and P. De Boeck (2015). Detection of differential item functioning using the lasso approach. *Journal of Educational and Behavioral Statistics*, 111–135.

Maij-de Meij, A. M., H. Kelderman, and H. van der Flier (2008). Fitting a mixture item response theory model to personality questionnaire data: Characterizing latent classes and investigating possibilities for improving prediction. *Applied Psychological Measurement 32*(8), 611–631.

Mair, P. (2018). *Modern psychometrics with R.* Springer.

Mair, P. and R. Hatzinger (2007a). CML based estimation of extended Rasch models with the eRm package in R. *Psychology Science 49*(1), 26.

Mair, P. and R. Hatzinger (2007b). Extended Rasch modeling: The eRm package for the application of IRT models in R. *Journal of Statistical Software 20*, 1–20.

Mair, P., R. Hatzinger, M. J. Maier, T. Rusch, and M. P. Mair (2016). Package eRm. *Vienna, Austria: R Foundation.*

Mair, P., Y. Rosseel, and K. Gruber (2023). Cran task view: Psychometric models and methods. version 2023-07-18.

Mantel, N. and W. Haenszel (1959). Statistical aspects of the analysis of data from retrospective studies of disease. *Journal of the National Cancer Institute 22*(4), 719–748.

Manuguerra, M., G. Z. Heller, and J. Ma (2020). Continuous ordinal regression for analysis of visual analogue scales: The R package ordinalcont. *Journal of Statistical Software 96*, 1–25.

Masters, G. N. (1982). A Rasch model for partial credit scoring. *Psychometrika 47*, 149–174.

Masters, G. N. (2018). Partial credit model. In W. van der Linden (Ed.), *Handbook of item response theory*, pp. 109–126.

Masters, G. N. and B. Wright (1984). The essential process in a family of measurement models. *Psychometrika 49*, 529–544.

Mauerer, I. and G. Tutz (2024). An ordinal item response model for understanding attitudes. Technical report, SSRN.

McFadden, D. (1973). Conditional logit analysis of qualitative choice behaviour. In P. Zarembka (Ed.), *Frontiers in Econometrics.* New York: Academic Press.

McNeish, D. (2024). Practical implications of sum scores being psychometrics? greatest accomplishment. *psychometrika 89*(4), 1148–1169.

McNeish, D. and M. G. Wolf (2020). Thinking twice about sum scores. *Behavior research methods 52*, 2287–2305.

Meier, L., S. van de Geer, and P. Bühlmann (2008). The group lasso for logistic regression. *Journal of the Royal Statistical Society, Series B 70*, 53–71.

Meiser, T., H. Plieninger, and M. Henninger (2019). IRTree models with ordinal and multidimensional decision nodes for response styles and trait-based rating responses. *British Journal of Mathematical and Statistical Psychology 72*, 501–516.

Messick, S. (1989). Meaning and values in test validation: The science and ethics of assessment. *Educational researcher 18*(2), 5–11.

Messick, S. (1991). Psychology and methodology of response styles. In R. E. Snow and D. E. Wiley (Eds.), *Improving inquiry in social science: A volume in honor of Lee J. Cronbach*, pp. 161–200. Hillsdale, NJ: Lawrence Erlbaum Associates, Publishers.

Michell, J. (2014). *An introduction to the logic of psychological measurement*. Psychology Press.

Millsap, R. and H. Everson (1993). Methodology review: Statistical approaches for assessing measurement bias. *Applied Psychological Measurement 17*(4), 297–334.

Minusa, S., T. Matsumura, K. Esaki, Y. Shao, C. Yoshimura, and H. Mizuno (2024). Response style characterization for repeated measures using the visual analogue scale. *arXiv preprint* arXiv:2403.10136.

Mokken, R. J. (1971). *A theory and procedure of scale analysis*. Berlin: Walter de Gruyter.

Molenaar, D., M. Cúri, and J. L. Bazán (2022). Zero and one inflated item response theory models for bounded continuous data. *Journal of Educational and Behavioral Statistics 47*(6), 693–735.

Molenaar, I. W. (1995). Estimation of item parameters. In G. Fischer and I. Molenaar (Eds.), *Rasch models: Foundations, recent developments, and applications*, pp. 39–51. Springer.

Moors, G. (2010). Ranking the ratings: A latent-class regression model to control for overall agreement in opinion research. *International Journal of Public Opinion Research 22*(1), 93–119.

Morgan, J. N. and J. A. Sonquist (1963). Problems in the analysis of survey data, and a proposal. *Journal of the American Statistical Association 58*, 415–435.

Muraki, E. (1992). A generalized partial credit model: Application of an EM algorithm. *Applied Psychological Measurement 16*(2), 159–176.

Muraki, E. (1997). A generalized partial credit model. *Handbook of modern item response theory*, 153–164.

Muraki, E. and M. Muraki (2016). Generalized partial credit model. In W. Van der Linden (Ed.), *Handbook of modern item response theory*, pp. 127–136. Chapman and Hall/CRC.

Noel, Y. and B. Dauvier (2007). A beta item response model for continuous bounded responses. *Applied Psychological Measurement 31*(1), 47–73.

Novick, M. R. (1966). The axioms and principal results of classical test theory. *Journal of mathematical psychology 3*(1), 1–18.

Nyquist, H. (1991). Restricted estimation of generalized linear models. *Applied Statistics 40*, 133–141.

Oelker, M. and G. Tutz (2017). A uniform framework for the combination of penalties in generalized structured models. *Advances in Data Analysis and Classification 11*, 97–120.

Ogasawara, H. (2001). Least squares estimation of item response theory linking coefficients. *Applied Psychological Measurement 25*(4), 373–383.

Osterlind, S. and H. Everson (2009). *Differential item functioning*, Volume 161. Sage Publications, Inc.

Oswald, F. and K. Schell (2010). Developing and scaling personality measures: Thurstone was right?but so far, Likert was not wrong. *3*(4), 481–484.

Peyhardi, J., C. Trottier, and Y. Guédon (2016). Partitioned conditional generalized linear models for categorical responses. *Statistical Modelling 16*(4), 297–321.

Pfanzagl, J. (1994). On item parameter estimation in certain latent trait models. In G. H. Fischer and D. Laming (Eds.), *Contributions to mathematical psychology, psychometrics, and methodology*, pp. 249–263. Springer.

Plieninger, H. (2016). Mountain or molehill? a simulation study on the impact of response styles. *Educational and Psychological Measurement 77*, 32–53.

Plieninger, H. (2021). Developing and applying IR-tree models: Guidelines, caveats, and an extension to multiple groups. *Organizational Research Methods 24*(3), 654–670.
Quinlan, J. R. (1986). Industion of decision trees. *Machine Learning 1*, 81–106.
Quinlan, J. R. (1993). *Programs for Machine Learning*. San Francisco: Morgan Kaufmann PublisherInc.
R Core Team (2021). *R: A Language and Environment for Statistical Computing*. Vienna, Austria: R Foundation for Statistical Computing.
Raju, N. S. (1988). The area between two item characteristic curves. *Psychometrika 53*(4), 495–502.
Rasch, G. (1960). *Probabilistic models for some intelligence and attainment tests*. Danish Institute for Educational Research.
Rasch, G. (1961a). On general laws and the meaning of measurement in psychology. In J. Neyman (Ed.), *Proceedings of the Fourth Berkeley Symposium on Mathematical Statistics and Probability*, Berkeley.
Rasch, G. (1961b). On general laws and the meaning of measurement in psychology. In *Proceedings of the fourth Berkeley symposium on mathematical statistics and probability*, Volume 4, pp. 321–333.
Rasch, G. (1966). An item analysis which takes individual differences into account. *British journal of mathematical and statistical psychology 19*(1), 49–57.
Rasch, G. (1977). On specific objectivity. an attempt at formalizing the request for generality and validity of scientific statements in symposium on scientific objectivity, vedbaek, mau 14-16, 1976. *Danish Year-Book of Philosophy Kobenhavn 14*, 58–94.
Rattinger, H., S. Roßteutscher, R. Schmitt-Beck, B. Weßels, and C. Wolf (2014). Pre-election cross section (GLES 2013). *GESIS Data Archive, Cologne ZA5700 Data file Version 2.0.0*.
Raykov, T. (1997). Estimation of composite reliability for congeneric measures. *Applied Psychological Measurement 21*(2), 173–184.
Reif, K. and A. Melich (1995). *Euro-barometer 38.1: Consumer protection and perceptions of science and technology, November 1992*. Inter-university Consortium for Political and Social Research.
Reise, S. P. (2010). Thurstone might have been right about attitudes, but drasgow, chernyshenko, and stark fail to make the case for personality. *Industrial and Organizational Psychology 3*(4), 485–488.
Rizopoulos, D. (2006). ltm: An r package for latent variable modelling and item response theory analyses. *Journal of Statistical Software 17*(5), 1–25.
Roberts, J. S., J. R. Donoghue, and J. E. Laughlin (2000). A general item response theory model for unfolding unidimensional polytomous responses. *Applied Psychological Measurement 24*(1), 3–32.
Robitzsch, A. (2024). Pairwise likelihood estimation of the 2PL model with locally dependent item responses. *Applied Sciences 14*(6), 2652.
Robitzsch, A., T. Kiefer, and M. Wu (2024). *TAM: Test Analysis Modules*. R package version 4.2-21.
Rogers, H. (2005). Differential item functioning. *Encyclopedia of Statistics in Behavioral Science*.
Roskam, E. E. and P. G. Jansen (1984). A new derivation of the Rasch model. *20*, 293–307.
Rost, J. (1990). Rasch models in latent classes: An integration of two approaches to item analysis. *Applied Psychological Measurement 14*(3), 271–282.
Ruppert, D., M. P. Wand, and R. J. Carroll (2009). Semiparametric regression during 2003 – 2007. *Electronic Journal of Statistics 3*, 1193–1256.
Samejima, F. (1969). Estimation of latent ability using a response pattern of graded scores. *Psychometrika monograph supplement*.
Samejima, F. (1973). Homogeneous case of the continuous response model. *Psychometrika 38*(2), 203–219.
Samejima, F. (1995). Acceleration model in the heterogeneous case of the general graded response model. *Psychometrika 60*(4), 549–572.
Samejima, F. (2000). Logistic positive exponent family of models: Virtue of asymmetric item characteristic curves. *Psychometrika 65*(3), 319–335.

Samejima, F. (2016). Graded response model. In W. Van der Linden (Ed.), *Handbook of item response theory*, pp. 95–108.
Schauberger, G. (2022). *PCMRS: Model Response Styles in Partial Credit Models*. R package version 0.1-4.
Schauberger, G. (2024a). *GPCMlasso: Differential Item Functioning in Generalized Partial Credit Models*. R package version 0.1-7.
Schauberger, G. (2024b). *MultOrdRS: Model Multivariate Ordinal Responses Including Response Styles*. R package version 0.1-3.
Schauberger, G. and P. Mair (2020). A regularization approach for the detection of differential item functioning in generalized partial credit models. *Behavior Research Methods 52*(1), 279–294.
Schauberger, G. and G. Tutz (2016). Detection of differential item functioning in Rasch models by boosting techniques. *British Journal of Mathematical and Statistical Psychology, to appear 69*, 80–103.
Schauberger, G. and G. Tutz (2022). Multivariate ordinal random effects models including subject and group specific response style effects. *Statistical Modelling*, 409–429.
Schwarz, G. (1978). Estimating the dimension of a model. *Annals of Statistics 6*, 461–464.
Searle, S., G. Casella, and C. McCulloch (1992). *Variance Components*. New York: Wiley.
Segerstedt, B. (1992). On ordinary ridge regression in generalized linear models. *Communications in Statistics – Theory and Methods 21*, 2227–2246.
Shmueli, G., T. P. Minka, J. B. Kadane, S. Borle, and P. Boatwright (2005). A useful distribution for fitting discrete data: revival of the conway–maxwell–poisson distribution. *Journal of the Royal Statistical Society: Series C (Applied Statistics) 54*(1), 127–142.
Sijtsma, K., J. L. Ellis, and D. Borsboom (2024). Recognize the value of the sum score, psychometrics? greatest accomplishment. *Psychometrika 89*(1), 84–117.
Sijtsma, K. and B. T. Hemker (2000). A taxonomy of irt models for ordering persons and items using simple sum scores. *Journal of Educational and Behavioral Statistics 25*(4), 391–415.
Sijtsma, K. and I. W. Molenaar (2002). *Introduction to nonparametric item response theory*, Volume 5. Sage.
Sijtsma, K. and I. W. Molenaar (2016). Mokken models. In W. Van der Linden (Ed.), *Handbook of item response theory*, pp. 303–321. Chapman and Hall/CRC.
Silvia, P. J., R. E. Beaty, and E. C. Nusbaum (2013). Verbal fluency and creativity: General and specific contributions of broad retrieval ability (gr) factors to divergent thinking. *Intelligence 41*(5), 328–340.
Soares, T., F. Gonçalves, and D. Gamerman (2009). An integrated bayesian model for dif analysis. *Journal of Educational and Behavioral Statistics 34*(3), 348–377.
Spearman, C. (1904). The proof and measurement of association between two things. *American Journal of Psychology 15*(1), 72–101.
Stevens, S. S. (1946). On the theory of scales of measurement. *Science 103*, 667–680.
Stevens, S. S. (1951). Mathematics, measurement and psychophysics. In S. S. Stevens (Ed.), *Handbook of Experimental Psychology*. New York: Wiley.
Strobl, C., J. Kopf, and A. Zeileis (2015). Rasch trees: A new method for detecting differential item functioning in the Rasch model model. *Psychometrika 80*(2), 289–316.
Strobl, C., J. Malley, and G. Tutz (2009). An Introduction to Recursive Partitioning: Rationale, Application and Characteristics of Classification and Regression Trees, Bagging and Random Forests. *Psychological Methods 14*, 323–348.
Suppes, P. and D. H. Krantz (2007). *Foundations of measurement: Geometrical, threshold, and probabilistic representations*, Volume 2. Courier Corporation.
Süss, H.-M., K. Oberauer, W. W. Wittmann, O. Wilhelm, and R. Schulze (2002). Working-memory capacity explains reasoning ability and a little bit more. *Intelligence 30*(3), 261–288.
Swaminathan, H., R. K. Hambleton, and H. J. Rogers (2006). Assessing the fit of item response theory models. *Handbook of statistics 26*, 683–718.
Swaminathan, H. and H. J. Rogers (1990). Detecting differential item functioning using logistic regression procedures. *Journal of Educational measurement 27*(4), 361–370.

Takane, Y. and J. De Leeuw (1987). On the relationship between item response theory and factor analysis of discretized variables. *Psychometrika 52*(3), 393–408.

Tendeiro, J. N. and S. Castro-Alvarez (2019). Ggum: An r package for fitting the generalized graded unfolding model. *Applied psychological measurement 43*(2), 172–173.

Thissen, D. (1982). Marginal maximum likelihood estimation for the one-parameter logistic model. *Psychometrika 47*(2), 175–186.

Thissen, D. and L. Cai (2016). Nominal categories model. In W. Van der Linden (Ed.), *Handbook of Modern Item Response Theory*, pp. 51–73. Chapman and Hall/CRC.

Thissen, D. and L. Steinberg (1986). A taxonomy of item response models. *Psychometrika 51*(4), 567–577.

Thissen, D. and L. Steinberg (2010). Using item response theory to disentangle constructs at different levels of generality.

Thissen, D. and L. Steinberg (2020). An intellectual history of parametric item response theory models in the twentieth century. *Chinese/English Journal of Educational Measurement and Evaluation 1*(1), 5.

Thissen, D., L. Steinberg, and H. Wainer (1993). Detection of differential item functioning using the parameters of item response models. *Differential item functioning*, 67–113.

Thissen, D., L. Steinberg, and H. Wainer (2013). Use of item response theory in the study of group differences in trace lines. In H. Wainer and H. Braun (Eds.), *Test validity*, pp. 147–169. Routledge.

Thissen-Roe, A. and D. Thissen (2013). A two-decision model for responses to Likert-type items. *Journal of Educational and Behavioral Statistics 38*(5), 522–547.

Thurstone, L. L. (1925). A method of scaling psychological and educational tests. *Journal of educational psychology 16*(7), 433.

Thurstone, L. L. (1927). A law of comparative judgement. *Psychological Review 34*, 273–286.

Thurstone, L. L. (1928). Attitudes can be measured. *American journal of Sociology 33*(4), 529–554.

Thurstone, L. L. (1929). Theory of attitude measurement. *Psychological review 36*(3), 222.

Tibshirani, R. (1996). Regression shrinkage and selection via the lasso. *Journal of the Royal Statistical Society B 58*, 267–288.

Tutz, G. (1989). Compound regression models for categorical ordinal data. *Biometrical Journal 31*, 259–272.

Tutz, G. (1990). Sequential item response models with an ordered response. *British Journal of Statistical and Mathematical Psychology 43*, 39–55.

Tutz, G. (2012). *Regression for Categorical Data.* Cambridge University Press.

Tutz, G. (2020a). Hierarchical models for the analysis of Likert scales in regression and item response analysis. *International Statistical Review* (89), 18–35.

Tutz, G. (2020b). On the structure of ordered latent trait models. *Journal of Mathematical Psychology 96*, 102346.

Tutz, G. (2022a). A flexible item response models for count data: the count thresholds model. *Applied Psychological Measurement* (46), 643–661.

Tutz, G. (2022b). Item response thresholds models: A general class of models for varying types of items. *Psychometrika* (87), 1238–1269.

Tutz, G. (2023). Unidimensionality in Rasch models: Efficient item selection and hierarchical clustering methods based on marginal estimates. http://arxiv.org/abs/2309.00553.

Tutz, G. (2024a). Invariance of comparisons: Separation of item and person parameters beyond Rasch models. *Journal of Mathematical Psychology*, 102876.

Tutz, G. (2024b). A latent trait approach to the measurement of physical fitness. *Measurement: Interdisciplinary Research and Perspectives*, 1–14.

Tutz, G. (2025). A general framework for random effects models for binary, ordinal, count type and continuous dependent variables including variable selection. *Statistical Modelling*.

Tutz, G. and M. Berger (2016a). Item-focussed trees for the identification of items in differential item functioning. *Psychometrika*, 727–750.

Tutz, G. and M. Berger (2016b). Response styles in rating scales - simultaneous modelling of content-related effects and the tendency to middle or extreme categories. *Journal of Educational and Behavioral Statistics 41*, 239–268.

Tutz, G. and C. Draxler (2019). A common framework for classical and tree-based item response models including extended hierarchically structured models. Technical Report 227, Department of Statistics LMU, Ludwig-Maximilians-Universität München.

Tutz, G. and J. Gertheiss (2016). Regularized regression for categorical data (with discussion). *Statistical Modelling 16*, 161–200.

Tutz, G. and P. Jordan (2024). Latent trait item response models for continuous responses. *Journal of Educational and Behavioral Statistics 49*(4), 499–532.

Tutz, G. and G. Schauberger (2015). A penalty approach to differential item functioning in Rasch models. *Psychometrika 80*, 21–43.

Tutz, G. and G. Schauberger (2020). Uncertainty in latent trait models. *Applied Psychological Measurement 44*(6), 447–464.

Tutz, G., G. Schauberger, and M. Berger (2018). Response styles in the partial credit model. *Applied Psychological Measurement 42*, 407–427.

Van den Noortgate, W. and P. De Boeck (2005). Assessing and explaining differential item functioning using logistic mixed models. *Journal of Educational and Behavioral Statistics 30*(4), 443–464.

Van der Linden, W. (2016a). *Handbook of Item Response Theory*. Chapman and Hall/CRC.

Van der Linden, W. (2016b). Introduction. In W. Van der Linden (Ed.), *Handbook of modern item response theory*, pp. 1–9. Chapman and Hall/CRC.

Van der Linden, W. J. (2016). Lognormal response-time model. In *Handbook of Item Response Theory, Volume One*, pp. 289–310. Chapman and Hall/CRC.

Van Rosmalen, J., H. Van Herk, and P. Groenen (2010). Identifying response styles: A latent-class bilinear multinomial logit model. *Journal of Marketing Research 47*(1), 157–172.

Van Vaerenbergh, Y. and T. D. Thomas (2013). Response styles in survey research: A literature review of antecedents, consequences, and remedies. *International Journal of Public Opinion Research 25*(2), 195–217.

Verguts, T. and P. De Boeck (2001). Some Mantel-Haenszel tests of Rasch model assumptions. *British Journal of Mathematical and Statistical Psychology 54*(1), 21–37.

Verhelst, N. D., C. Glas, and H. De Vries (1997). A steps model to analyze partial credit. In *Handbook of modern item response theory*, pp. 123–138. Springer.

Verhelst, N. D. and C. A. Glas (1995). The one parameter logistic model. In G. Fischer and I. Molenaar (Eds.), *Rasch models*, pp. 215–237. Springer.

von Davier, M. (2016). Rasch model. In W. van der Linden (Ed.), *Handbook of Item Response Theory, Volume One*, pp. 59–76. Chapman and Hall/CRC.

von Davier, M. and K. Yamamoto (2007). Mixture-distribution and hybrid Rasch models. In *Multivariate and mixture distribution Rasch models*, pp. 99–115. Springer.

Wand, M. P. (2000). A comparison of regression spline smoothing procedures. *Computational Statistics 15*, 443–462.

Wang, T. and L. Zeng (1998). Item parameter estimation for a continuous response model using an EM algorithm. *Applied Psychological Measurement 22*(4), 333–344.

Warm, T. A. (1989). Weighted likelihood estimation of ability in item response theory. *Psychometrika 54*(3), 427–450.

Wetzel, E. and C. H. Carstensen (2017). Multidimensional modeling of traits and response styles. *European Journal of Psychological Assessment* (33), 352–364.

Wood, S. N. (2006a). On confidence intervals for generalized additive models based on penalized regression splines. *Australian & New Zealand Journal of Statistics 48*, 445–464.

Wood, S. N. (2006b). Thin plate regression splines. *Journal of the Royal Statistical Society, Series B 65*, 95–114.

Yellott, J. I. (1977). The relationship between Luce's choice axiom, Thurstone's theory of comparative judgement, and the double exponential distribution. *Journal of Mathematical Psychology 15*, 109–144.

Yuan, M. and Y. Lin (2006). Model selection and estimation in regression with grouped variables. *Journal of the Royal Statistical Society B 68*, 49–67.

Zhao, P., G. Rocha, and B. Yu (2009). The composite absolute penalties family for grouped and hierarchical variable selection. *Annals of Statistics 37*, 3468–3497.

Zhou, X. (2019). Hierarchical item response models for analyzing public opinion. *Political Analysis 27*(4), 481–502.

ZPID (2013). PsychData of the Leibniz Institute for Psychology Information ZPID. *Trier: Center for Research Data in Psychology*.

Zumbo, B. (1999). A handbook on the theory and methods of differential item functioning. *Ottawa: National Defense Headquarters 160*.

Zwinderman, A. H. (1995). Pairwise parameter estimation in Rasch models. *Applied Psychological Measurement 19*(4), 369–375.

Index

G. Tutz, *A Short Guide to Item Response Theory Models*, Statistics for Social and Behavioral Sciences, https://doi.org/10.1007/978-3-031-87271-6

S

T

U

V

Author Index

G. Tutz, *A Short Guide to Item Response Theory Models*, Statistics for Social and Behavioral Sciences, https://doi.org/10.1007/978-3-031-87271-6

The manufacturer's authorised representative in the EU is Springer
Nature Customer Service Centre GmbH, Europaplatz 3, 69115 Heidelberg,
Germany. If you have any concerns regarding our products, please
contact ProductSafety@springernature.com

Printed and bound by CPI Group (UK) Ltd, Croydon, CR0 4YY
01/10/2026
02225960-0010